THE CARE AND MAINTENANCE OF HEAVY ***jets***...

By Christian Parker

This is NOT approved aircraft maintenance data. This book is intended as an orientation to some aspects of airline aircraft maintenance culture. It is not intended to be a reference in any way to actually performing aircraft maintenance.

The opinions expressed in this book are entirely the author's, and do not necessarily represent those of any corporation.

Published by OLYMPIC DISPATCH LLC
Port Townsend, Washington.

ISBN-10: 0-9815838-3-0
ISBN-13: 978-0-9815838-3-9

Library of Congress Control Number: 2008933184

1st edition printed 2008.
2nd edition first printing 2009, second printing 2016.

Printed in the United States of America

CONTENTS

Introduction

This book is ultimately about unions and their connection to socialism in America; which is ultimately about a two-centuries old, European atheist-inspired conspiracy to bring about a new world order. When the author was growing up in the 1970s through early 80s, he didn't give a hoot about these topics. In high school this was the most boring type of stuff imaginable. But after life starts to affect us—*then* we become interested…and when we start to scratch under the surface, what *will* we find?

The author grew up in a Pacific Northwest Scandinavian-dominated extended family that would produce successful business proprietors, engineers, two PhDs, a CEO, and a judge and Army general on one side, and then artists and musicians on the other side. The author's mother sent him to the little church right across the street in Scandia, Washington from a tender age and later to Christian private schools for junior high and high school. A love of history was fostered while growing up at grandpa and grandma's houses reading old editions of National Geographic magazine and listening to tales of the early immigrant's exploits in the northwest, sea stories, and heroic deeds in WWI and WWII. True to the stereotypical northwest political stance, there were a lot of liberal Democrats and good 'ole American pro-union supporters in the group, but not all. From this we have the seeds of a worldview!

I am not God's gift to the aviation community, but I have been working in heavy aircraft maintenance for most of my adult life now. First in the Air Force and then for a repair station that does the heavy aircraft maintenance for various airlines. In both cases it involved run-of-mill, transport aircraft. I also had a very brief stint at one of the world's largest airlines. And of course I have known many other aircraft mechanics that have worked for several major airlines, and I have had a lot of contact with people who have worked for the major aircraft manufacturers as well. I am not a Ph.D., an engineer, or a commercial pilot, but I have known some of them too. My perspective, my research venues, are from near the bottom: my job, trade magazines, and inside Intel. I have heard the rumors and seen the real evidences. Oh, and I am a very religious person, a Christian. I guess you could call me a Jesus freak.

This book was originally focused on improving airline aircraft maintenance, and that is the subject treated, but the problem is really much broader than that. This book takes to task the institutions in American industrial corporate system in general. The item of most concern is American jobs going overseas, and why. What can be done about it?

I had begun this book in the late 1990s, but it had languished until 2007, when I was hired by Boeing to work as a flightline AMT for the new 787 Dreamliner project. It was watching the flailing, incredibly inefficient 787 manufacturing process that finally pushed me over the edge to actually finishing the book. Something had to be done in American aerospace, and American industry.

Boeing had assembled quite a team from across the industry for the Delivery Center flightline and the flight test organization. People from United Airlines, Northwest, American, TWA, Delta, and others. When shown manuscripts of this work, or discussed the topics with them, most felt I had pretty much hit the nail on the head.

Between my military service and airliner work, I did a lot of personal study in history and religion. I became very interested in that word *anthropology*: the study of man. Why are things and people the way they are. I think of myself as an amateur historian and anthropologist. Western civilization consumed my time for a couple of years. I spent hundreds of entire days reading in coffee shops, the University of Washington libraries, and various other libraries. I developed quite a used-book store habit. For a while, in my late twenties/early thirties, I seriously wanted to be an archeologist or anthropologist. I had a bit of an Indiana Jones complex you could say. Maybe it was all those *National Geographic's* I had read since a young boy (yes, I actually read them, at least a little).

This book is written from the perspective of an historical and anthropologic study. The career of the author is given as exhibit A. Whether the conclusions given are qualified or not, the reader will have to assess for themselves. I suppose the reader may be thankful that the author is not a product of the pervasive ideologies of the modern university system, as if he were this work would likely be unreadable and nauseating.

A true academic would spend their time in the study of original sources, perhaps in foreign languages, and live in a sterile environment fraternizing with other academics. From time to time a study would be performed, a study from the perspective of an outsider looking in. The author has witnessed such studies and has seen their ineffectiveness first hand. Conversely the author is a mostly self-educated researcher, who has lived in the world of labor from his beginning. If we suppose that human nature is the same now as it ever was, this experience is relevant to all ages. All that is needed then is knowledge of history in order to see what has been tried before and to what result. Included in this edition is a bibliography to show some basis of that research into history, anthropology, and technology.

I am very interested in family history and became a bit of a genealogist, culminating in my writing a comprehensive family history. I found out that I am related to Captain John Parker of the Lexington Militia, 1775, who took the message from Paul Revere that the British were coming, and acted upon it. I found that I was also related to a famous early nineteenth century Boston abolitionist, the Reverend Theodore Parker. So it seems that I have some rabble-rousers in my past. This makes me feel more of an American, like I can't just sit by while things go to pot.

The author has one grandfather and two great-grandfathers who were Chief Engineers on steam ships. In researching my family history I came to be familiar with the story of the American Merchant Marine. I found there to be a lot of similarity to the American airline industry. Indeed, there seems to be more similarities than differences in the various transportation industries.

It is interesting to note that the International Association of Machinists (IAM) union started with the railroads, and has represented members of the maritime industry, the aviation industry, and others. The Machinists union represents most of the hourly workers at Boeing, and at one time mechanics at many of the major airlines.

The author's great-grandfather was a member of the IAM when he was just starting out as a marine engineer circa 1900, at a junior engine room position called an oiler. When he graduated on to third assistant engineer, he joined the Marine Engineers Beneficial Association (MEBA). That fact made me wonder why it was necessary to switch unions. The Machinists union was apparently for the lower ranking people. Why was that, I wondered? If the union was simply there to protect people, why switch to another? Why the need to have two? One should have been just as good as the next, but apparently they were not.

Several of the people of my family are pro-union, indeed my grandfather Charles Parker, who was my hero, led a strike of the marine engineers on the Washington State Ferry system in the late 1960s as a business agent for the MEBA at that time. But I have come to question the need for some of America's unions, as well as the liberal politics that are behind them.

My dad, Charles Parker jr., entered an apprenticeship at the Puget Sound Naval Shipyard in Bremerton, Washington as an electrician in the 1960s. He was an energetic worker and had a sharp mind, and was a foreman by age twenty-eight. Following this he took an inspector job with the office of the Supervisor of Shipbuilding (SupShip), which oversaw and approved contracts the Navy awarded to private shipyards. The group at SupShip were a fraternal chummy lot, and entertained each other frequently. It was in this atmosphere that the author grew up, hearing of the demise of the private sector U.S. ship repair business. In the early 1990s they all got laid off—the domestic ship repair had gone away. The only real work then being done was in the actual government naval yards. The U.S. Merchant Marine, the commercial cargo and tanker ships, was effectively dead by this time—it had been driven overseas to foreign ownership—and the pervasive opinion was that unions and government actions of the post World War II democrat controlled congress had killed the entire private sector of the industry—and with it the jobs.

Chapter 1

The Nature of the Business

Have you ever wondered about the people and the world of airline aircraft maintenance? If it's small aircraft you're talking about, there are numerous books and resources at your disposal. But if it is the world of airlines you are interested in, it is not too easy to find a book on airline heavy turbine aircraft maintenance and the world therein. The author could not find too much when he tried to find something on heavy aircraft maintenance when going to A & P School in the early 1990s. That is why I decided to write this book, because I thought you might like to know. The airline and large aircraft world is afflicted with a "big brother is watching" kind of mentality. People are worried about "them," and what "they" are looking at. Everyone is afraid of his or her shadow it seems sometimes.

This book is for anyone who is thinking about airline aircraft maintenance as a career, and anyone else who is just curious about all that. When I was a young man the airlines were still in their golden age heyday, anything and anyone to do with airlines was and were lofty. Today it's not so glamorous. Now it's down and dirty; it is transportation, just like the bus, train, and ship.

This book is also a bit of a political treatise, as the airline maintenance world is sheer paperwork, politics, and intrigue. For example, no one could ever convince me that TWA flight 800, perhaps the greatest aviation incident of recent memory, was not shot down or blown up. It sure sounds like it was shot down by all the eyewitness accounts to that effect. The government excavated the remains of the plane from the sea, pieced it all together, and then formed their conclusions. They concluded that no bomb or missile was the cause of the disaster. They think that a combination of the fuel being heated up with the air conditioning packs running while on the ramp at JFK, and then sparking in the fuel tank from bad wiring or something set it off. If that's the case, God help us all. Better up the insurance policy.

When Boeing is testing a new aircraft, they will take it and run it in the deserts of Australia or somewhere similar. They will run the AC packs for days on end in the sweltering heat. The author learned this from a former Boeing engineer many years ago. They will "proof" test the aircraft in everyway possible. So when I saw newspapers of the time talking about AC packs, which are under the center fuel tank, heating up the fuel because the aircraft sat on the ground for a long time, I began to believe in conspiracy theories. The business about wiring in the tank making sparks and so forth is just a bunch of governmental aviation world spin designed to throw people off. There is no air in the tank of a jet liner, only fuel and fuel vapors, neither of which can burn without air. I have seen evidence to this effect with my own eyes. You can have sparks in the tanks from static and lightning strikes. The powers that be are fully aware of this. And besides, kerosene doesn't light off that easy. No, this is misinformation in the military sense. The engineers do a fairly good job of designing these airplanes, and most accidents are the result of some human action, not chance, not scientific anomalies, not bad luck. The people didn't latch the latch, didn't put the bolt in, didn't put grease on the jackscrew, did put the box of oxygen generators in the cargo pit, left the rig pin in the elevator quadrant, didn't remember to put the flaps in the right position, fired the missile at the jet, etc. Of course there are a few scientific anomalies like the weather phenomenon of microburst, but in general, the causes for things seem to be action-reaction. Just because something could happen, doesn't mean that it actually does in real life. There is a greater power guiding events.

I have what you call an A & P license, which stands for Airframe and Powerplant license, which is granted by the FAA. The FAA is of course the governmental agency that oversees the aviation community, the Federal Aviation Administration. It was also supposed to be called an AMT(T) license, Aircraft Maintenance Technician(Transport), but the FAA apparently changed its mind about that. You can get one by going to a school that has an aviation maintenance program for two years, or you can also get one by having prior aircraft maintenance experience such as from the military. In either case you will then have to take some tests to actually get the license. In the school they will teach some useful stuff, and some not very useful stuff like how to put on the dope and fabric of a WWI era plane, like the Red Baron's. In fact, a lot of the school will be centered around 1930's technology airplanes. This is because that is the way the FAA wants it at this present time. Their reasons are?

In aviation, you have "general aviation" which covers all of the small and pleasure airplanes, corporate aviation which is the emerging large fleet of business jets, and airline or transport aviation which is all about money.

It helps to understand a couple of things when talking about the development of the airline industry. First, the military made it all happen by providing the initial money for research and development and then buying a ton of airplanes, thus giving the aircraft manufacturers the money and know-how needed to build airliners. Thus the airline and heavy aircraft industries are steeped in a government/military "big" sort of mentality. And second, the airlines have essentially taken over from the railroads and merchant marine, neither one of which really hauls passengers any more, at least not in the United States. Those ocean liners of old were about getting people from A to B first, and providing a luxury vacation second. That is why they used to have such a premium on speed. The luxury vacation was an outgrowth of the "just getting there" business. The luxury ocean liners of today have very little in common with the *Titanic* and ships of the past, now they are simply floating resorts. So even though the airlines are not very old, historically speaking, they have continued on in a transportation industry that is older and has many old traditions.

There is hardly an aircraft mechanic that has not romanticized about being a P-51 pilot over the skies of Europe in World War II, or a Spitfire jockey over the Battle of Britain. These are the romantic visions that drove many to aviation in the first place. In those days, a pilot was like a bird that flew vicariously through his machine. The airplane was an extension of his hands, feet, and senses. The pilot flew the airplane himself, through direct mechanical control. There was not much in the way of high tech toys on them. The basic instruments, and some guns. And the pilot was not only there to fly the plane, but to fight, he was a warrior, the mounted knight. The military pilot is a completely different job than a civilian airline pilot.

The airliners were fairly basic in the beginning, but they had a few toys. The autopilot was on the large planes at an early date. An airliner is like a passenger or merchant ship, the pilot is the captain, and the auto-pilot is the helmsman of an airliner. The captain and autopilot's job is to keep the wheels pointed towards the earth. Like on a ship, the captain has the issues of navigation, weather, the rest of the crew and passengers, and the security of the cargo. The pilot used to be there to land and take off, but these days the autopilot can do that as well. These days, there are even two autopilots, in case one has a problem. The pilot is the officer in charge, and is there in case of an emergency.

The first big transports and cargo planes had a flight crew of several: The captain, first officer, flight engineer, navigator, and radioman. The flight engineer was an approximation of the engineer of a ship. He was there to monitor the aircraft's systems and engines. On some of the old reciprocating (propeller) planes, he had direct control of the engine throttles. On some of the larger planes, the engineer could even crawl through a passageway in the huge, fat, wings to gain access to the piston engines. They really had to keep an eye on things because mechanical failures were common, and in-flight engine shutdowns were a regular thing.

The job of flying was shared by the pilots, navigator, and radioman. But the march of technology began to displace some of the aircrew. First the radioman was lost to progress, then the navigator. The microchip really made its presence known in the airliners, and the flight engineer position was eliminated, and even the pilots were to be found looking over their shoulder. The computers were taking over. In the future, and even now, it will be possible to have planes without pilots at all. The military already has them. The current generation of piloted military fighters might possibly be the last. As for me, I like the idea of a real live human pilot up there looking out for his or her own welfare. Computers have already flown one or two planes into the ground, against their human pilot's will of course. But, the computer is seen as a better pilot, and so aircraft manufacturers will continue to make planes that think for themselves. Indeed, a modern airliner can be seen as like an organism, it has a skeleton, joints, muscles, tendons, skin, respiratory system, circulatory system, nervous system, and brain. Body, wings, and feet, the great "air beast." In some aircraft today, a pilot is merely like the rider of a horse, a horse that knows its way home. The pilot pulls the winged horse's reigns to say right, left, up, down, but the horse does the motion itself. It is "buffered." This means that the pilot cannot take the aircraft out of its controllable flight envelope. This makes it safer. Unless you are on an unfortunate A-320, and the computer augurs you into the earth.

Well, the role of the mechanic has not really gotten any easier over the evolution of the airliner. True, the old airplanes were more labor intensive in some ways, the engines were a lot less reliable in the early days, and needed to be changed, and otherwise worked on often. The engine failures have dropped off dramatically with technological advancement. That is why we have planes with only two engines these days. And engineers will improve this and that and the other thing. But from the maintenance standpoint, with improvements come more sophisticated technology and complexity. And unlike an organism, a plane cannot heal itself.

Do you know that aircraft maintenance was originally classified by the government as unskilled labor?! Does that make you feel good knowing that the airplane you and a hundred or more other people have trusted their lives in is maintained by unskilled workers? Probably not! It would not make me feel very good either if I truly believed that. But I know better, and I know that aviation maintenance professionals are very skilled workers, or at least the core workers that every organization has, which tend to make everything happen even in spite of the system. There is still much improvement needed in the aviation maintenance industry, but government, corporate, and union forces seem to work to keep the industry standards low.

An airline mechanic is ideally a jack of many trades: A mechanic for both engines and airframes, structural repairman, machinist, electrician, electronics technician, inspector, technology expert, system diagnostician, tooling engineer, aviation paralegal, researcher, operator of all manner of light and heavy equipment, including various sizes of forklifts, cranes, and cherry pickers, and aircraft ground operator. But in reality the average mechanic will become an expert in only a few of these disciplines. Some mechanics barely rise above "blue water technician."(that's a joke) A few mechanics will master most or all of these areas in a career.

At an airline the jet maintenance falls basically into two categories, flight line, and hangar. There are two worlds represented here. The flight line is part of the "operations" environment. That is the daily flying business right along with the ramp service, flight attendants, and pilots on the gates at the airports. The hangar is the realm of aircraft maintenance entirely; it is the maintenance world, nothing but mechanics and a few engineers. The hangar work is the overhaul and preventive maintenance work. The hangar might not even be at a major airport.

The flight line is where the emergencies happen and get fixed, and also a bit of everyday preventive maintenance takes place like lubing the various bearings and checking the tires. The flight line mechanics working the gates at airports are like firemen or EMTs, or maybe even the emergency room doctor. They spend a lot of time playing cards in a waiting area, but when they are needed they have to be on it.

The airplane has a maintenance logbook that the pilots, flight attendants, or other mechanics can enter discrepancies that are encountered. After a flight, the line mechanics will come to the aircraft and give it a look over for safety of flight issues. They will check the logbook and address any items found in there. The airplane has an FAA approved minimum equipment list, which lists equipment that is essential for flight dispatch, and items that may be deferred. If the problem is not an emergency, or one that does not have to be fixed before the next flight, they might opt to defer the problem to be fixed at the next scheduled maintenance visit when they will have more time to address the problem. They tend to exercise this option quite a bit.

The line mechanics are usually qualified to run the airplane's engines, and taxi the aircraft around the airport as needed. In fact, they operate the aircraft in every way but actually flying it. I knew one mechanic who worked for a foreign airline that allowed high speed taxiing for troubleshooting purposes (not many of them do). Many times he took a 747, flaps up and spoilers up to dump the lift, and ran it down the runway over 150 miles an hour to check for wheel shimmy or something.

That is just the beginning of course, on top of operating the aircraft; they have to figure out what is wrong with it. They need to be familiar with all the aircraft's systems so they can determine what to do quickly when a problem arises. Fix or defer, what to do? The airline world obviously runs on a busy schedule, and so they do not always have the luxury of time.

Traditionally the line mechanic is considered the more skilled because he has to fix whatever is wrong quickly, or at least determine what needs to be done quickly. He might have to change an engine or a wing flap, a black box or some module, an air cycle machine or a light bulb. But most of the time it is tires and brakes, and minor stuff. The line mechanic is often the more experienced mechanic. Well, that is how it is supposed to be anyway. At many airlines, they will have specialists that work primarily with the aircraft's electronic equipment, called avionics. There is a trend towards integrating the avionics technician and A & P, however.

In the hangar the preventive maintenance takes place, so that emergencies happen less on the flight line and in the air. The work in here is steadier and a lot more comprehensive. Traditionally, the hangar mechanic is the lower seniority and less experienced mechanic. This is where the future line mechanic "learns the plane." The hangar could be compared to a hospital for jets, and an intensive care unit. At the same time it could be compared to a traditional shipyard, because of its size and scope.

The hangar mechanic is also often more specialized than the line mechanic. He might work almost exclusively on structures, or electrical wiring. Sometimes the flight line might request specialists or experts from the hangar to come out and fix a specific problem. Also in the category of hangar maintenance are the people who overhaul engines and other components. These are general stereotypes of the maintenance world, and of course exceptions apply.

The airline business is the latest incarnation of an old and established transportation industry, and is the successor to the sea ship or maritime business that came before it in many ways, commonly referred to as the merchant marine[1]. It is a natural evolution of the sea trade. Navigation, terminology, passenger and freight processing methods, corporate structures, labor structures and labor problems, and the most advanced use of technology aside from the military. Steamship line to airline, shipyard to aircraft manufacturer. At least in the United States, the airplane all but killed the passenger sea trade, and the passenger railroad trade, at the same time yielding to the jet age where everyone could afford travel.

Airplanes are built similar to the wooden sailing ships of old, and very much like the riveted iron and then steel ships of the nineteenth and early twentieth centuries. It is no wonder that Seattle Scandinavian boat people built the biggest airplane company in the world. They are made of metal instead of wood of course, but they are similar in that they comprise of a skin covering a skeleton, and made of thousands of small parts held together by rivet or rivet-like fasteners. Airliners are sometimes referred to as

---

[1] For a take on the history of the U.S. merchant marine see *On a Single Stack Steamer: plying Northwest waters* by the author, which is a prequel to this work.

ships. An airframe can be referred to as a *hull*. A Boeing airliner has a *keel* beam at its bottom. Below the airplane's outer skin are frames, ribs, stringers, longerons, intercostals, and all are held together with rivets and bolts of various types. The skins of the wings are called *planks*.

Just like the mechanism to steer a ship's rudder, the airplane uses cables and pulleys to steer its rudder, as well as all its other steering parts. In the operation of a ship we refer to its axis of movement as pitch, roll, and yaw. The same is for the airplane. When referring to parts of the aircraft, we say inboard, outboard, forward, aft. A Boeing plane has a "water line" as one of its measurement parameters; something is either so many inches above, or below, the water line.

In navigation, both plane and ship use the same type of navigation data. Both measure speed in "knots." An airplane has a green navigation light for right (starboard), and a red one for left (port), just like on a ship. Just as a ship pulls into a port, an airplane lands at an air-"port." An airplane's passenger entry doors are usually on the "port' side. Each mode works with the government and customs in the same way, and uses documents of a similar nature, such as a manifest of passengers…and on and on….

The American merchant marine has a long and storied history. In the days of early America, the era of the sailing ship, the ships of New England grew to prominence in the world's merchant fleets. Particularly notable were the whale ships, which were to be found everywhere on the world's oceans. By the 1840s the American merchant marine had taken a place equal to Europe's fleets in the world. This had been achieved by American ingenuity and perseverance alone, and not by government subsidy or help. But then something new arrived on the world's oceans—the steamship. Eager to maintain a prominent place at sea, the United States government would subsidize a new steam merchant fleet. This would ultimately have the opposite effect of what had been desired. It created a weak fleet that failed.

Then came the destructive effects of the American Civil War, which decimated the American merchant marine. Following the war, the U.S. merchant marine fell behind the European powers at sea. America did not embrace the steamship following the war, but instead relied on the tried and true sail ship of her glory days. Conditions on the sailing ships grew harsher, giving birth to the American west coast "Shanghai" and an oppressive system of obtaining sailors at west coast ports.

The advent of the unions.

In the land we know as Europe…since the last days of the Romans…the pagan hordes of the north had gradually succumbed to the teachings of the Christ…their kings ruled by God-granted right…the church government, often corrupt, grew to wield almost limitless power…but the true faith burned bright in the hearts of the people—giving the church and state their power. Early Christian Europe had an order: those who work, the nobility, and the church. In the 1600s a purification of the church took place, known as the Reformation…

Trailing the Christian Reformation in Europe, the intellectual movement known as the Enlightenment spans the 1700s. The Enlightenment would be heralded as reason and science over the traditional sway of the church and state; towards a world not dominated by the church (God). This would foster the great social upheavals of the

French Revolution, the Terror, and Napoleon—events that would change Europe forever. Emerging from this chaotic period in the early to mid 1800s, atheist intellectuals began to contemplate a new world order—Socialism—a world without God. Karl Marx is of course the icon and most famous of these with his communism. There was a congress of like-minded socialists who founded an organization called the International Workingmen's Association (IWA). This is the beginning of the modern labor union movement and the communist/anarchist/syndicalist movements. In 1880s San Francisco, the IWA movement surfaced and fostered sailor's unions and others. Communist radicals would be part of these new unions throughout their history. Thus began a century of struggle between ship owners and seamen, management and labor. There would be huge strikes following WWI, in the 1930s, and following WWII.

Also over the next century, the government would get involved with building and operating merchant vessels, originally to "bring back" America as a maritime power. Of course government owned industry is a hallmark of socialism. Each time the government got involved in the ship business, however, it upset the applecart and ultimately wrought havoc.

Born in the last decade of the nineteenth century, based out of Seattle, was the Alaska Steamship Company. Their market focus was on hauling passengers and freight from Alaska to the west coast ports of the United States in the wake of the Klondike Gold Rush. Their main business was passengers, but they carried cargo also. They did a thriving business through the first part of the twentieth century, but in the 1950s passenger ridership fell, and costs, particularly union labor costs, skyrocketed. Of course the main reason passenger ridership fell was due to the rise of air travel. After many decades of popular service, Alaska Steam went out of business.

Today, filling the very same role, we have Alaska Airlines serving the west coast of the United States, mainly carrying passengers, but also cargo. Their focus was originally on Alaska trade. Alaska Airlines hauls a lot of fish down from Alaska.

Even though sea born passenger service waned, however, sea born freight transportation grew and grew. The era of inter-modal containerization saw an explosion in the size of ships. World ship lines were alive and well, and continue to be to this day.

American industry seems to be having a bit of trouble in the late twentieth and early twenty-first century keeping labor in-house. America used to have a large merchant fleet. It is gone now, overseas. We used to build a lot of ships here. Now we don't, that too, overseas. We used to make almost all the cars we drove, televisions, etc. Much of that too is now done overseas.

The author is fond of outboard boat motors; to work on them, run them, collect them. America used to absolutely dominate the world outboard motor industry. Like many things, we invented them just after the turn of the twentieth century. This was a time when American farm boys and tinkers were inventing all sorts of things. Edwardian America was a time when at least some Americans were still self reliant, industrious, and entrepreneurial. The airplane was invented by mechanics in this same era, not formally educated mechanical engineers like we might have expected.

Early American manufacturing industry was centered in New England and the east coast. In the twentieth century the Great Lakes region became synonymous with American manufacturing. One of the firms of this region, one that would eventually be known as Outboard Marine Corporation (OMC) would grow in the 20th century to be the world's largest maker of recreational boat products, best known for their lines of outboard motors *Johnson* and *Evinrude*, although they were diversified in other areas as well. This was an iconic American company from America's heartland. Ole Evinrude, son of Norwegian immigrants and a Wisconsin farm boy with some engineering smarts is credited with inventing and producing the first successful mass-produced outboard motors. Ole never had a formal education beyond the third grade, but he was self-taught in mathematics, as a mechanical engineer, pattern maker, machinist, steel worker, and businessman. His son Ralph would see his first two years of college, then join the family business never to return to school.

In the first years of the twentieth century Ole wanted to be a carmaker and had several failed business attempts. Along the way he would marry his life's love Bess, who had a head for business and together they would be unstoppable. He conceived the outboard motor concept in 1906, which in 1909 would prove to be an instant success. Ole would see his company grow, sell out, and reenter the industry. In 1929 Outboard Motors Corporation was formed by the merger of three companies, and with Ole Evinrude as president. The company was based in Milwaukee and had the lines of outboards *Evinrude, Elto,* and *Lockwood.* Ole died in 1934 and Ralph took the reins until he retired in 1982. The Depression took its toll on the industry, and in 1935 a weakened Johnson Motor Company was merged into the firm along with its modern factory in Waukegan, Illinois. In 1936 the firm took the name Outboard Marine and Manufacturing Company, with the lines of outboards *Evinrude, Elto,* and *Johnson.* The company was now headquartered in Waukegan and had diversified its operations. Outboards were the primary product lines. The company would grow to have facilities all over the world and employ thousands.

But OMC was far more than just its leaders; the company was populated by industrious farm boys turned mechanics from the American heartland like Ole Evinrude. God-fearing people of traditional values. This is what it ultimately takes to have a successful manufacturing operation. A quality focused efficient team. Executives might make light of that, but that is terribly difficult to achieve, and when it is achieved—that firm wins.

But in the twentieth century the American people did not trust their management for their well-being, and the unionizing spirit was high. OMC was no exception. The Great Lakes region, like America's other industrial regions, was a hotbed of industry, Democrat politics, and union activity. Many Americans believed unions were "American" and democratic. But what many didn't realize was that the labor union was a direct outgrowth of communism, a European invention devised by atheists and not really American at all, and with the stated intention of establishing a new world order.

Even though a great many of the union members did not hold radical left-wing socialist political ideals, the union was perceived as bringing-home-the-bacon, and that trumped all other concerns.

The original outboard was a single cylinder unit of 1.5hp. Then came a horizontally opposed two-cylinder. None of them were over 4hp until 1925. The early motors employed a gearbox having forward only, and either a knob on the flywheel for a starter, or later a rope start flywheel. Motors employed a tiller handle for steering. Motor size started to grow in the fast-paced years of the late 1920s, when they grew to 25hp, and the four-cylinder horizontally opposed "quad" engine was introduced. Then the Depression hit and the market shrank, but OMC hung in there and continued to innovate. In the 1930s OMC produced motors as much as 40hp, but the smaller cheaper motors would dominate the market. In the 1930s the motors received partial powerhead cowling, optional rewind starters and the larger motors had optional electric starting. The alternate firing twin was introduced, an engine design that would dominate the line in postwar years.

During WWII OMC made aviation war supplies, and introduced 50hp motors for military use. Post war, the pre-war designs were held over for a few years, but new designs were forthcoming and OMC kicked into high gear with alternate-firing vertical inline twins with reversing gearboxes in integral one-piece lower units, full powerhead cowling, rewind starters as standard, full remote controls on some models, remote larger capacity fuel tanks, and electric starting on some models. All models were now fully painted in fashionable livery. Americans increasingly wanted more power and in 1958 the first V4 was introduced at 50hp, which by 1967 would grow to 100hp. The sterndrive inboard/outboard system was introduced and refined. OMC became the largest maker of outboard motors in the world because of its products and support. They built the motor that would get you there and back. When American quality was king, OMC was a king of American quality. Always at the forefront of technology, in 1967 high-technology breakerless ignition was employed for the first time. And

OMC continued to grow and innovate from there. OMC surged ahead better than ever in the 1970s with the most technically advanced motors in the world when American carmakers were struggling with the energy crisis, the EPA, and poor quality. In the 1970s Japanese car sales were killing domestic carmakers, but in the outboard motor market there was almost no foreign competition, and OMC led the domestic pack unquestioningly. There was significant foreign competition in the inboard/outboard sterndrive market from Volvo Penta, however.

The EPA was close on their heels, however, as OMC's manufacturing plant in Waukegan, Illinois just north of Chicago was targeted for a massive cleanup. A number of AFL unions served OMC's labor force, and the typical contract issue labor problems would crop up. There was a big strike in 1977.

Japanese competition joined the fray, chiefly Yamaha and Honda, and others as well. Despite the field becoming larger, in the 1980s things continued to go well. There were some labor troubles at a plant in Georgia. The next generation of motors were being engineered.

The company had until this time been run by executives grounded in engineering or a genuine passion for the industry. In 1990 the last of these retired. At this time OMC still commanded 60% of the market. A new breed of mediocre leadership would take over from here. More labor problems.

In the 1990s new fuel injection technology was being developed in concert with a German firm to deal with environmental standards. While the industry was moving to four-stroke engines to deal with this, OMC was pioneering fuel injection technology for two-stroke engines to keep their power-to-weight-ratio competitive, while achieving improved fuel economy and cleaner emissions. Four-stroke engines by comparison are much larger and heavier than two-strokes.

Like so many American companies, OMC would outsource its manufacturing abroad to control costs—to the significant loss of quality in some cases. In 1995 there was a strike followed by continued labor unrest. Strikes typically cost a company economically and in reputation. One can argue whether or not a union workforce costs more in lost productivity or not, or if greater quality compensates. But when a company outsources, its management is doing so for reduced costs, however you want to look at it. They are stating by the act that they are not satisfied with what they have. They have to outsource to "stay competitive...." Labor will blame management, and management will blame labor productivity, quality, and union activity; when perhaps the blame is more evenly distributed. It is no secret that American corporations were looking to "beat" the unions in the late 20th century, but if so why? Did they have good reason?

The late '90s saw turmoil in the company's leadership and the situation became dire fast. In 1996 the company reported 1.1B in revenue, but following that the drop was precipitous. In 1997 OMC came under the ownership of Greenmarine Acqisition Corporation/Geenmarine Holdings LLC/Greenway Partners/Soros Fund Management LLC, or simply "Greenmarine" which was a group of investors that had no experience running a manufacturing company, and of which America's puppet-master George Soros was involved. In 1998 they began closure of the Waukegan and Milwaukee manufacturing plants. Rumors of financial woes. Executives came and went. The outsourcing scheme proved to be a disaster, and dealers had to deal with quality issues right out of the crate. The new fuel injection technology proved to have serious teething problems. Japanese competition, and domestic Mercury were right on OMC's heels. The Milwaukee plant was closed in 2000. After nearly a century of market leadership, the American icon OMC struggled and eventually succumbed to bankruptcy in late 2000. OMC laid off almost all of its workers. The outboard motor division was sold to a Canadian company. In 2002 the Waukegan factory was abandoned and eventually mowed down, and the EPA cleanup of the site would go on for many years. It seemed to represent American industry in general—invented the product, reigned in its market for most of the century, and collapsed of? Unfortunately OMC was not in the "too big to fail" category.

OMC's last downfall was in large part due to its laying off the bulk of their legacy workforce and outsourcing to a foreign country, with disastrous effects on quality. This reflects poor management strategic decision-making. Causal contributing factors are decades of union activity and the frustrating never-ending battle. The entire system failed. Result: Everyone lost.

The Canadian company that bought Evinrude and Johnson with the new technology would go on to work out the fuel injection problems and make the cleanest running and best power-to-weight-ratio motors in the world....

The airline industry got going in the 1930s with the airplanes big enough to carry passengers like the DC-3. This was also a big union organizing time and so the airline mechanics got on the union bandwagon right off. Unions like the International Association of Machinists and Aerospace Workers (IAM) quickly gained sway. Government regulation of the airline industry was instituted in this early period as well with the Civil Aeronautics Board (CAB), being formed in 1938.

The jet age was when the airline industry really took off, however, and the jets became the dominant transportation service for passenger travel. The dawn of the jet age saw the sunset on the steamships and railroads for passenger service, at least in the United States. The jets were ultra "high tech" and awe inspiring. In the early jet days, an airline mechanic had to have had military jet experience, and the expectations were high. Working for the airlines was very glamorous. Of course being a flight attendant was ultra hip.

The sixties and seventies were heydays for the airline world. Then came airline deregulation in 1978. Suddenly the airlines had to compete more like the other businesses around them. By the mid-eighties the airline world was completely deregulated and CAB was dissolved. The airline world got more business like and down and dirty. Some airlines would fail in this environment, Braniff was the first. Airlines suddenly began looking for ways to cut costs to better compete in a competitor rich environment. The airlines started subcontracting some activities that they had traditionally accomplished themselves. In the aircraft maintenance world there began a trend to subcontract out the heavy and expensive airframe overhaul work, and even some of the routine daily work. Most of these operations were non-union. This did not sit well with the unions, who tend to rely on numbers of members, and "solidarity."

During this same period the parent industry of commercial aviation, the U.S. merchant marine was dying—the entire industry—and going overseas due to a number of factors, not the least of which was its union labor and a century of struggle with the asset owning entities.

This was also to be a period of airline worker unrest. Airline mechanics were represented by various unions, but one of the most dominant unions was the IAM, at one time representing mechanics at Pan Am, Eastern, United, Northwest, Trans World Airways, Continental, Western, and several others. In the post deregulation upheaval, the traditional unions such as the IAM would lose membership. Some of the airlines they represented went out of business such as Pan Am and Eastern. Continental would simply vote them out and have no union. TWA would be absorbed by American in a takeover, TWA's mechanics being absorbed by the Transport Workers Union (TWU). Western was absorbed by Delta, which always had non-union mechanics.

Other airlines would switch to a new "craft" union called the Aircraft Mechanics Fraternal Association (AMFA). The mechanics at some airlines had come to think that they were not being represented to the best effect by the traditional unions. They felt the union tended to lump different groups together when negotiating benefits. The mechanics, cleaners, ramp service, warehouse workers, fleet service, and others were grouped together, and pay was averaged among these groups. The mechanics wanted a union more like the Airline Pilots Association, one that was more professional and just for them. AMFA stepped in to fill this need. In the years following the turn of the millennium the IAM represented only a couple of the smaller airlines, and by the middle of the first decade AMFA had represented a majority of the important airlines that were union represented, such as United, Northwest, and Southwest.

But, results were not perceived as forthcoming at some airlines, and the much-pruned mechanics at United voted in the Teamsters union in early 2008. In the same period, more union-mechanic airlines went out of business. The heavy maintenance at Northwest got outsourced overseas, and then the historic merger of Northwest Airlines being absorbed by Delta in 2009 eliminated still more union jobs. The old-school union aspect of the industry seemed to be flailing wildly in pain.

There are several airlines that are non-union, at least for mechanics, such as Delta, Continental, Jet Blue, and Skywest. And then there are the mostly non-union repair stations.

In hard times the pressure to cut costs continually dogged the airline maintenance world, and some of the overhaul work started going overseas, particularly after the new millennium. Will the American aviation industry eventually suffer the same fate as the other American industries that have gone "over there?"

What kind of person becomes an aircraft mechanic?

A "real" mechanic starts learning his or her trade at a very young age. It starts with a curiosity of what makes things work. Similar to an interest of what makes animals and organisms live. A lot of people do not really realize that a mechanic starts learning his trade as a young child, playing with toys that require some sort of logic to put together, i.e... the round peg in the round hole. Some have this curiosity and some do not. Mechanics that were young boys after World War II in the 1950s, 60s and 70s grew up without computers to take up all their time. As young boys they played with *Legos*, *Erector Sets*, model electric trains, and they built plastic model cars, planes, and ships. Maybe they even had a chemistry set or an electronic assembly toy. Maybe model rockets, or remote control-line airplanes, the kind that flew in circles with the two strings that controlled their elevation. In dealing with all these toys, real world frustrations of how to make them actually work were an everyday reality for these youngsters. There were real world engineering problems involved. Like how to put the fire cracker in the plane or tank just right to make it blow up and melt in just the right way to make it realistic and fun. There were maintenance issues involved, like the need to clean the electric train tracks of their oxidation, so the train would go again. And in many of these things, these boys had to learn a basic truth of mechanical things; they don't always work precisely as they are engineered to. They often need a little "massaging." And these boys had television to sap up much of their time—the World War II generation really had to use their imaginations to amuse themselves.

It is generally agreed by experienced mechanics, that to be a good mechanic, one generally must possess the "knack," or mechanical inclination. A good mechanic typically grows up taking all sorts of things apart to see how they work and are quite fascinated by this. Thus they are developing mechanical dexterity and training in mechanical anatomy for their career from childhood. This is combined in later years with a practical knowledge of engineering principles and methods to the extent that makes a good mechanic/technician. A one or two year program taken in adulthood is *not* a good substitute for this lifelong training. Very often a person with no mechanical experience at all who takes a vocational/college program is not much better a mechanic/technician than before taking the program, though they can boast more knowledge and likely pass a test. If they are a better communicator than the good mechanic, they may very well seem to be a better mechanic when talking to someone, though they are not and this would readily be evident by a comparison of their work.

When one looks at where good mechanics come from, we think of the farmer, and the farm community, where skills required for the business are respected and coveted throughout the entire community. It might be seen that such skills grow out of a culture, a culture that values them.

The rural farm boy often seems to make the better mechanic, and worker in general for that matter, because of the necessity of doing the farm chores, fixing the thing that leaks, the thing that doesn't work, making the thing work better so less physical effort is needed. The farm boy is the one who took apart the lawn mower (and got in trouble!). But one day, many a farm boy learned to put the lawn mower back together, or maybe his mini-bike. And then it was the tractor, the truck, and etc.

The fisherman is also an inventive and astute mechanic. They say that necessity is the mother of invention. The marine trades were the original "high-tech" trades of mankind. Ships and boats employed the greatest use of technology. The sea has always been a demanding environment, and making a living from the sea has been a difficult and arduous task throughout time. The construction of ships and boats has evolved through time by the shipwright/mechanic through trial and error methods, without using engineering drawings, and was taken to a high degree. The methods employed by whalers and fishermen, the tackle and equipment involved, represented some of the greatest everyday employment of technology in ancient and historic times.

The people who built the greatest airplane company in the world, Boeing, were predominantly Scandinavian immigrants. They came from relatively poor countries where fish and fishing were a prominent part of life. Norway in particular was known as a land of water and rocks. They were used to having to work hard for a little, and employing the use of technology in building boats and fishing tackle systems and whaling methods. This was combined with a pronounced old world European Christian era moral and work ethic and an American atmosphere of unbridled opportunity and willingness to embrace new industry. The maritime prowess of Scandinavia is well known from the Viking era.

It has been the observation of the author in his aviation career that many Pacific island Asian men are above average mechanics, particularly Japanese, Polynesians, Moros, and others. Would there be a particular reason for this? All these peoples came from areas where their ancestors were involved in maritime trades, fish being so important a part of the diet of these regions. So we have the employment of maritime technology in the building of boats and fishing tackles in a large part of the population. Somehow this tendency towards the application of technology is handed down through their culture. When presented with a modern industrial technology, the jump was often readily made. Some would point out that these peoples, like the ancient Europeans, were warlike and had well developed weapons, but weapon manufacture was more specialized and involves a smaller part of the culture, while the everyday business of getting food or trading with neighboring villages involves nearly everyone.

The Makah Native American tribe of Cape Flattery at the northwest corner of what is now the United States employed a complex system of buoys and harpoons to hunt whales, which was very similar to the system used by European and American whalers of the nineteenth century. The only thing the Makah lacked was the European style whale ship itself, which was unnecessary to them as they were always able to get their required catch of whales right off the coast near the village, and their long canoes were sufficient.

The farm boy and fisherman are also close to nature, which seems to be the mechanical engineering inspiration for all. Leonardo daVinci observed nature to invent all sorts of things. God is a good act to follow. I remember one pleasant afternoon in the early 1990s, walking on Kingston beach near the ferry landing when the tide was out. I looked down and saw a severed crab claw lying in the sand. The side of claw had been pecked open by the Sea Gull, obviously to eat the meat inside. But all the working parts of the crab claw were there. As I studied the little thing, it dawned on me that I understood how the thing was constructed. The muscle pulled primarily one way only. We learn from the medical industry that muscles contract, or get shorter. The large muscle in the claw was oriented for pulling on the lower attach point for the upper claw assembly. This gave a lot of clamping force to close the claw, but almost none to open it. The lower half of the claw was a heavy structure, and only the upper claw moved to provide the clamping. The lower claw was then both half the "claw," but also a housing for the other claw actuation mechanisms, which explains its large size. The inside of the lower claw was neatly clearanced for the movement of the lower lever of the upper claw assembly. The upper claw pivot joints were nice and beefy "ball" joints designed to take a lot of stress, but with very low friction. The muscle pulled at a point as far from the fulcrum axis as possible, taking advantage of the maximum moment arm, inside the lower claw structure to gain maximum power for the downward clamping action of the claw. The muscle itself, which is soft, was attached to a large fan like piece of cartilage. If the muscle were attached to a small point, it would rip off. This fan shaped piece of cartilage then focused its energy force through a small, very tough, somewhat flexible piece of material that connected to the upper claw, lower attach point. The biological equipment inside the claw was sensitive to contamination, so dynamic seals were provided at the movement joints, and little hairs, like brushes, kept the sand from lodging between the pieces. This was a mil-spec advanced technology structure! This was an aerospace structure! It amazed me that I could look at something devised by God and figure it out so easily (I wonder where God got His engineering approved?). But I was used to looking at mechanical things, designed by engineers, who had originally been inspired by someone like daVinci.

There seem to be a few predominant types of people that become aircraft mechanics. There are the hot-rodder/auto mechanic/motorcycle mechanic types that are interested in the technology. I have always been around these, especially when I worked on jet engines down in California. The hot-rodders definitely make some of the best mechanics. Then there are the pilots and aviation enthusiast types. These have less mechanical skills, but often more aviation knowledge. They know the performance details of every airplane ever made. If they stick with it and don't get discouraged by their lack of mechanical ability, they usually pan out. Some of them who grew up with dad's airplane or something are actually excellent mechanics right from the get go. The thing about many general aviation types is that they are extremely prideful. The general aviation type tends to be a purist. Almost every one sees himself as an Orville or Wilbur Wright. In general aviation they do have to be engineer, and mechanic both very often. But the simple fact is, there is just not that much too the small prop planes. They are fairly simple by comparison with an airliner. And, most of the small planes do not get that much work done to them, by comparison, either. But, the older general aviation mechanic is known to be a major know-it-all. They spend all their time talking, and not enough doing. They believe they are far superior to military and airline types, and, maybe sometimes they're right. One thing is for sure though, they tend to make less money that the airline mechanic. On the other side, the general aviation mechanic is a little more likely to be a pilot, and that is a good quality of any A & P mechanic. It shows professionalism. A knowledge of the operations environment is very helpful to understanding many things.

Then there are the people who go to A & P School because they are looking for a good job. They do not have much in the way of mechanical skills, and not much in the way of aviation knowledge. This group tends to struggle. Some of them will work out, some will not. But, like in anything else, if they want it bad enough they will attain some level of proficiency.

There are a lot of people who learn to work on aircraft in the military. Most of these of course are from the Air Force or Navy. But they also come from the Army, Marines, and Coast Guard. People are of course unique, but we are shaped to some extent by the culture in which we live. Each military has a culture of its own. You would think the Air Force would be the best of these, because it is all about aerospace, but only a small minority of Air Force mechanics learn a great deal of skills in their military. The first reason is that the Air Force is very specialized in their career fields; most will learn only a part of a plane. And the second reason is that a person in the Air Force is often a leader after a few years, and is only a worker for a short period for that reason. The military is built around having a person for only that four-year enlistment, and therefore does not take a long-term career approach. The Air Force is also very anal and particular, and doesn't like mistakes. They put a high emphasis on quality. They therefore do not encourage people to be go-getters with their work, but more careful. They have some go-getters of course, but these are a minority. The Air Force people do learn corporate politics better than anybody else, however, and they have the best living conditions.

The Navy guys are get-it-done scrappy types, but can be a little on the hack side of things. Not everybody of course, but the nature of their mission dictates a make-do-with-what-you-have approach. They do have more accidents. They are used to working with less and needing greater results, in a crammed environment. They have my respect. The Navy guy tends to have broader and more practical experience. You will work in the Navy. The other services seem to do quite well with their aviation units as well, even though their resources are less. They tend to be self reliant and capable. Military experience by itself will not typically get you an airline job. You will have to get an A & P license first in most cases. Military experience will, however, often get you a job at a repair station, where you can get your license, more and broader experience, then an airline job. Airlines more and more desire repair station experience, because it tends to be broader based experience. Military experience, repair station experience, an A & P license, an FCC General Radiotelephone Operator license, and maybe a college degree are a winning combination. For Avionics, military experience rules. This is one area where the military does very well. But airlines will likely want you to have an A & P license as well.

One thing about the auto mechanic types is that they do not always play well with others. They tend to be proud and independent. An auto mechanic is used to working by himself. One thing the military does is teach a person how to work in an organization, to be a part of a team. And heavy turbine aviation is a team effort. The ideal mechanic, however, is one that is able to work independently, yet still be able to function in the "society" of the maintenance hangar.

But if you do not want to go into the military, you certainly do not have to, to get into aviation. The military and the civilian airlines are not the same. The military has unique opportunities that only it can fulfill. In the military you can have the greatest situation, or a really lousy one, it is kind of luck of the draw. People who got into flying jobs in the military generally have good memories of the service, wouldn't trade it for anything. Maintenance people are more mixed in their views. I am glad I went into the military myself, best thing I ever did as a young man. But many military aircraft are very different than the airliners. The hardware is the same though, the nuts, bolts, safety-wire, hydraulic tubing, wiring, etc, and so there is at least some commonality.

Many airlines will hire people strait out of A & P School. And you will certainly get a job at a repair station out of A & P school, which you can then leapfrog from to an airline. Or you can work general aviation out of A & P school.

The A & P school problem

One big problem in the aviation maintenance industry is that, very often, people that go to A & P School to learn how to work on airplanes do not graduate being able to do that. They pass the tests, and do all the schoolwork, but they do not seem to learn the fundamentals of mechanic work. They are not "mechanically inclined," or have "the knack." They got up one day and said, "I think I will be an aircraft mechanic." They did some research, found out about the FAA approved A & P schools, and signed up for the two-year program. And the A & P license has often been called a "license to learn." And perhaps it should be, but people should graduate a little more qualified. And the people who have no aptitude for the trade should find out before they get all the way through school. A few of the schools are very good, but far more are not.

On the other side of the coin are the farm boys and mechanical types that grew up doing things, and know how to turn a wrench, but do not see the point of doing the study and getting the license. These people are almost equally vexing. Some mechanics need to have some book smarts, because not all problems are simple, and jets aren't getting any simpler. Aviation needs the mechanical types that are willing to do some study.

The real answer to the problem is to make an industry wide shift. Instead of going to an A & P school that just wants to crank out graduates, there should be an industry wide apprenticeship program that combines school and "real" hands on training. The U.S. Merchant Marine had a program like this that started in the 1930s. The government got involved because there was deemed to be a general deficiency in the training and quality of merchant marine officers. Before this time, many merchant marine officers learned everything they knew on the job, without much formal schooling. The U.S. fleet was expanding at that time, and steps were put in place to make a better system.

The training program of the Merchant Marine Commission was thought to be a success in those early days. These were the mariners that carried the goods across the oceans in their Liberty ships and Victory ships during World War II. The program was an industry/government collaboration. It was administered through Kings Point Merchant Marine Academy of New York. There were extension campuses around the country. This was a program to train the officers only, but the concept seems sound to me. The officers were considered the "professionals." They made the choice to go to sea. There were two categories: deck officers, and the engineers. The engineers were responsible for the mechanicals of the ship, the "tech" people, and their ranks were broken up into four levels: Chief engineer, first assistant, second assistant, and third assistant. Below them were other workers, such as oilers and wipers to do the dirty work. An engineering officer had some control over his career advancement, in that he took tests for the different ranks.

The cadets had some entry requirements to be met, and then they attended some campus classroom studies. In general, the program was administered as a military school. But, instead of spending years at school, they went to sea as "apprentices," called "cadets," and performed much of their study there. They then alternated between sea and shore studies. It would be clear early on if a cadet did not have the aptitude for his job.

The government is involved now in the administration of A & P school curriculum. The FAA mandates what will be taught in U.S. A & P schools, and for how many hours, but the schools are private. So we are not that far away from the ideal. The system just needs to be combined with the airlines and repair stations. In fact, a school combined with an apprenticeship at a FAA Part 145 Repair Station would be the perfect combination. Then after the school period, the technician would learn more areas of the plane for a few more years, then go to work line maintenance and light checks at an airline. Or, they could stay at the repair station, and become an expert. I like the medical doctor/hospital model: medical school, internship, and residency.

A modern non-union Part 145 repair station is where a mechanic comes into a contact with a lot of work, and therefore the maximum exposure to learn the skills of the trade. Even a good mechanic that is "mechanically inclined," and has the "knack" takes about two years to become proficient with one segment of a modern jet liner's systems, such as the flight control systems, in this environment. And then there is still the avionics, the landing gear, and the engines. At a union type airline, the time is even longer, because the exposure to work is less. And if the person is not mechanically inclined they will take much longer or almost never to reach a level of independent proficiency. A bachelor's degree helps with research skills often, but will not make a good mechanic in and of itself. Like a doctor, instruction in the techniques and methods is what is lacking for these less mechanically adept people. Can these things be taught to people that did not growing up doing mechanical things? I think it is possible, but would take a careful analysis in this area, and a serious effort to teach people things that some would consider "obvious," or something that you just kind of "pick up."

At the present time, there is just one license to work on airplanes. Well, technically, it is actually three, general, airframe, and powerplant. It is generally thought of as one license, but it is actually two, airframe and powerplant. The general license is part of either of the other two. You get a general and airframe, a general and powerplant, or all three. The A & P license is oriented towards the general aviation community. Then there is an Inspector Authorization license (IA) that is also oriented towards the general aviation community. General aviation often does not have all the resources of the airline world. The general aviation mechanic is said to be the engineer and mechanic. They must use "approved" data to maintain the planes, but this still leaves them leeway.

The airline world has a lot more resources at hand. The airline world runs on paper. Everything is written down. The mechanic works to precise technical data provided by the manufacturer and airline engineering departments, which are all FAA approved. These data are very strict. When members of the FAA look at an airline's maintenance, they are looking at paper--at various documentation: Aircraft manuals, engineering orders, airline procedure manuals, and records. The mechanic is supposed to follow the data to the letter. The physical plane and the paper should always match. But just because they should doesn't mean they always do.

As with all types of manufacturer's repair data, aircraft manufacturer's repair data is quite strict, often times over strict. Most airlines maintenance manuals and tech data come from the manufacturer. The manufacturer has a master maintenance manual, and the airline takes the data from this that applies to their particular model aircraft. This is then submitted to the FAA and is approved.

The problem with generally over-strict repair criteria is that it tends to make the less informed mechanic a scoff-law. Many mechanics think that the maintenance manuals are ridiculously anal. The problem is however, sometimes it is best to be very strict, while in other repair situations the need to be exacting is far less critical. Judgment comes into play. The mechanic is not supposed to make these judgments, but they do and they will. No one wants to see an aviation incident, and that is not even on the table. The idea is that the maintenance will make it to the next inspection interval without incident.

When an engineering department and a maintenance department work well with each other, many or most of the engineering solutions to repair problems that come up are actually are proposed by the mechanics, and are then approved by engineering. This is of course is based upon physical observation and experience. This is quite legal. This is the best we have under the current system. At a few airlines with really crafty engineering departments, they have had their manuals and tech data modified to incorporate such "practical" maintenance and repair data. This is then FAA approved.

But airlines are not all created equal, and there are many different standards at play. Different FSDO's and different airlines. There are as many standards, it often seems, as there are FDSO's, or at least that is the perception. The difference between how much an airline's maintenance is scrutinized, is how much discipline is applied to verifying that the lofty written down standards are carried out. Is the airline held 100%

accountable? 75%? Maybe 50% of the written down standard? This is the problem. From the mechanics end, it appears as if the differences are political.

The engineers are in an office usually, removed from the aircraft. The engineer is not a hands-on mechanical person in most cases. They are often not of a "practical" nature. This is the culture our system has created. For these reasons it often takes a great deal of time to arrive at engineering solutions to problems, whether they be simple or complex. When a mechanic encounters a problem not specifically addressed by technical data, they are supposed to submit to engineering for an answer, and what to do. The trouble is that mechanics are often put in a position to make small maintenance and repair engineering decisions on a regular basis. This is due to constraints of time or money, like it or not. This is illegal, and quite a point of sensitivity, but it happens on a routine basis at all airlines, it is merely a matter of how often and how much. It is going to happen no matter how much the FAA throws a fit, so it would be a wiser approach to empower the mechanic slightly more, and regulate such behavior above board. The situation seems like that of Prohibition, in trying to do a good thing, the law ends up doing a bad thing, and creating problems. Because of this, it is not known what the true state of an airline's maintenance is based on a review of the paperwork alone.

Chapter II

An Individual Airline Mechanic Career

I started out in October 1983 going to Air Force basic training down in San Antonio, Texas. It was not overly difficult. Next I went to a school at Chanute Air Force Base (AFB) at Rantoul, Illinois. It was a school for jet engine maintenance and was only a couple of months long. The school was apparently just to teach us the basics. I will always remember that base because they had all kinds of cool old airplanes on museum duty. They had the only B-58 Hustler I have ever seen, awesome.

They trained us on some very old jet engines, even for that time. Most of our training was on the J-57, known in the civilian airline world as the Pratt & Whitney JT3. It was on the 707, the DC-8, some old fighters, and the original B-52 bomber. As I recall, we did not do too much hands on anyway, mostly classroom work. There was some interesting theory and technical information, but I came away thinking there should have been a little more to it. In the military they had a saying: J.E.E.P.

(Just Enough Education to Pass), and that about sums it up. I was a J.E.E.P.

In the first part of 1984 I was at my first active duty base, Travis AFB, California. It is nearby Fairfield and Vacaville in Northern California. The aircraft assigned to this base were the C-141 Starlifter, with TF-33 engines, and the C-5 Galaxy, with TF-39's. These were the aircraft of the Military Airlift Command (MAC). MAC was the cargo and passenger airline of the military. It probably had more in common to the airlines, than to the rest of the military.

I was assigned to the TF-33 engine shop. Again, I seemed to be on old engines. The TF-33 was basically the same thing as the civilian Pratt & Whitney JT3D, one step up from the JT3. Again this engine was found on military and civilian planes from the early sixties. The D in JT3D stands for ducted. This means that it had a bypass duct and fan. This made the engine a "low bypass" turbo fan. The regular JT3 had no bypass duct or fan and was thus a "turbo jet." I believe the JT3D was the first turbofan. Pretty much all the airliner engines are turbofans these days. The turbojets were the loudest things you ever heard, screeching thunder. The low bypass turbofans are not much quieter. Some airliners today have a close relative of the JT3D, the JT8D. The JT8D is off course very loud. In their day they were space age tech; but these are of course the piston steam engines of the jet age, underpowered and outdated by present standards, and way too loud.

Today we have the ultra high bypass turbofans. The TF-39 on the C-5 was the first of these. It was pretty much the same as the General Electric CF-6, like what was on the first 747's.

The TF-33 shop was pretty cool I guess. This was very much a civil service job; most of the key people and management of the engine shops were civilian government employees. They of course had a union, and the shop seemed a lot like a stereotypical union shop of the period. You had to fill up an eight-hour day somehow. They had some very good civilian mechanics in there, however. My first lead, a civilian, was a racecar engine builder on the side.

There were four or five crews on day shift, each crew had about five people. The crews were a mixture of military and civilian mechanics. Most of the crews on day shift had civilian leads. The job of one crew was to work on one engine, primarily. Secondarily, engines that had previously been built by that same crew, but were still being "shaken down" (inspected for minor discrepancies), were worked on. There was another crew to work on the thrust reversers and other sheet metal parts. The shop had its own tool room and parts room.

This shop really did a complete overhaul though. This was supposed to be field maintenance, which is an intermediate level of work. But they did virtually a "depot" level overhaul, the highest level of work. They would tear the engine apart down to nothing but pieces. When the Air Force was working with Pratt & Whitney to rewrite the intermediate maintenance manual for the TF-33, they used the experience of the civilian workers of this shop to write the book.

There were more people than work in the engine shops, so us new G.I.'s were usually fighting over something to do to look busy. Plus, we didn't know much, so there was only so much we could do. I remember that this could make for some really long days, because we had to stay there with the civilians whether there was work or not. If there was not much work, sometimes we would get farmed out to the TF-39 engine shop. When there was work, you spent all day every day with your crew, at a conversation distance. This of course could be interesting, a bunch of twenty year olds from different parts of the country bantering it out, and one or two older civilians standing aloof and amused.

When I was assigned to the TF-33 shop, I took a five-month correspondence course on jet engines, which was a lot more interesting and comprehensive than the initial school at Chanute. Still, it was centered around TF-33 type technology. At that time the more advanced fighter type engines were a different job code entirely.

Every year we had a public Air Expo at Travis, and I got to represent the TF-33 engine at this fair a couple of years. We had the engine displayed in the fair area and I got to answer any question I knew how to the passers by.

A jet engine works on a four-cycle system: intake, compression, combustion, and exhaust. In the aviation world there is a saying about jet engines: they suck, squeeze, bang, blow, and go.

Most large jet engines have two "spools," or compressor/turbine assemblies. These are usually referred to as the N1 for the first compressor/last turbine assy., and N2 for the second compressor/first turbine assy. Both the compressor and turbine assemblies have blades that move, called "blades," and blades that do not move called "stator vanes." They look the same. Between the compressor and turbine is the combustion chamber. On the outside of the engine is a gearbox for driving various things. It is driven via a shaft from the N2 compressor/turbine assembly. These days you can see this in a computer encyclopedia or on *Modern Marvels*.

The two rotating assemblies are supported by a series of roller bearings, and lubricated by a dry sump oiling system. Like the heart, arteries and veins in the body, the engines oiling system has an oil pump which supplies oil in tubes down the right side of the engine, called pressure lines, and then collects the oil in the bearing cavities and sends the oil back to the tank and pump down the left side of the engine in "scavenge" lines.

One of the rotating compressor/turbine shafts drives another power take-off shaft that drives a gearbox external to the engine. On this gearbox is mounted most of the engines vital mechanical sub-systems, like the organs of the body. The starter, fuel pump, oil pump, fuel control, generator and generator drive transmission, and various other do-hickies like one of the tachometer generators.

The physical structure of the engine is comprised of several "cases," like barrels, bolted end to end. Exhaust case, turbine case, combustion case, compressor case, and inlet guide vane case. You may have seen a picture of a jet engine somewhere and seen the rows of hundreds of bolts. Inside the cases are the bearings for the rotating shaft assemblies that the compressors and turbines ride on. These shaft assemblies are held together with some fancy spline & nut arrangements. Torquing these things is quite a deal, and a large part of the overhaul process. In fact, elaborate joining and torquing procedures are a large part of the basic engine buildup.

Like older cars this jet engine did not have electronic controls. It was old-fashioned mechanical, hydraulic, and pneumatic sensors and controls. It did have bimetal exhaust gas temperature (EGT) probes. These alumel and chromel probes make their own juice to power the EGT gauge in the flight deck. Of course these days a computer controls things on a jet engine, just like they do in your car. And just like your car, you do not have to do much "tuning" anymore.

The air can get bunched up between the two compressors and cause a "stall" or "surge", and so has to be regulated. The older engines had a bleed valve between the two compressors to regulate the airflow through the engine. The other method is to have variable stator vanes in the compressors to keep the airflow even. These are the principal differences in the jet engines. They are all "fuel injected." These days some engines have both variable stator vanes and variable bleed valves.

If you want to learn more about jet engine technology you will have to read another book, they are out there. Jet engines are cool.

As a jet engine ages, its operating EGT gets higher as a result of the engine's wear in the hot end. If it runs too hot, it will have to be removed and overhauled, or it could run the risk of a catastrophic failure. High vibration is also a reason to remove an engine. Ignoring high vibration can result in a catastrophic failure of the heavy turbine wheel flying out of the engine and possibly sawing through the fuselage of the airplane like a buzz saw. It has happened. On these old engines, excessive compressor stalls were a cause to remove an engine to see what's up. A compressor stall is kind of like a giant backfire out the carburetor. It's caused by improper airflow through the engines compressor. On a modern engine, any compressor stall is cause for engine removal and inspection. And, just like an old car, sometimes an engine was replaced because it was "burn'n oil," or too much oil that is, usually accompanied by high breather pressure (bearing cavity air pressure). A condition caused by bad carbon seals (which seal the oil in the bearing cavities), kind of a similar situation to a piston engine that has bad piston rings and is blowing oil. Sometimes an engine would have to be replaced due to a failure of the engine's physical structure; cracks, damage from ingestion of foreign objects (FOD), or shelled out turbines.

An engine would arrive at the shop with some problem. A visual inspection would be performed, especially in the turbine area. It would then be stripped of all its external cowlings, tubing, wiring, and most of its external parts. Then the cases would be separated, the turbines and turbine vanes removed, the combustion cans removed, the first stage turbine shaft removed, compressor disassembled, and etc.

All these parts would then be inspected and replaced as necessary. The first stage turbine vanes would usually be cooked pretty good, distorted, blackened, and eroded. Occasionally, a combustion-can might have a defect such as burning or cracks. Maybe the first stage turbine would have to be replaced. The bearing cavity carbon seals were often replaced.

Then the buildup would commence. This of course was done with greater caution than the tear down process, and took a lot longer. There was "tuning," or "blueprinting" involved in the buildup process, particularly with the turbine and turbine stator spacing, the turbine vane nozzle area adjustment, as well as a lot of very careful measuring, assembly and torquing processes.

After the insides of the engine were put together, the crew commenced on putting together the outsides of the engine: Gearbox and accessories, bleed valves and bleed components, engine mounts, ducts, the maze of tubing, wiring harness, ignition system, the fire detection loop, and cowlings.

After the engine was completely rebuilt, taking a crew on three shifts about two weeks, the engine would be taken out to the test cell and run extensively. Going out to the test cell on the Travis flightline was fun. It was way out in a park-like setting. Going to the test cell was a nice way to pass the day, it was called "rest cell." Most of the time the engine would run well, but sometimes there would be problems and the engine would have to go back to the shop. Sometimes the problems would be large and the engine would have to come back apart to have a carbon seal replaced or something. That was bad, and many would be displeased.

When the engine had passed its test cell run, it would then be placed in a storage hangar. And for a crew in the engine shop, the process would start over.

Like many, I really liked the technical aspects of my job, but I did not care for the military aspects as much. When I was at the engine shop almost a couple of years I had some trouble with the shop chief not liking me (I really don't know what his problem was) so I asked to leave for another shop. This was granted and I went to the C-141 Isochronal Inspection Dock, engine section. This was a medium maintenance visit for the aircraft. It was similar to the airline C-check. The plane was in the hangar for two or three days, and had mechanics crawling all over it. Anything you had to fix had to be done in that time, so when the plane arrived you had to be all over it.

First a thorough inspection, and at the same time, the oil, filters, and spark plugs all changed. Then, even before the inspection was done, you started fixing all the minor write-ups that were found. With any luck, that was about it: oil, filters, plugs, and few small discrepancies. Discrepancies like worn clamps on tubes and wires, cracked brackets and parts, cracked parts of cowlings, worn fireloops (fire detectors), worn thrust-reverser parts, rusty engine mounts, and cracked pneumatic ducts. For some strange reason or another we sometimes had to change a fuel control or oil pump or something difficult like that, but not often. We did change too many thrust-reversers, what a pain that was. On the flight line they changed a lot of generator and generator constant speed drive transmissions (CSD's), but we did not have to change them that often.

Often a first stage fan would have to be replaced because it had reached "high time" or "high cycles." This meant that the fan disk had been taken to takeoff thrust, and therefore maximum rpm, a certain number of times, and now had to come off and be non-destructively inspected to make sure it would not fly apart. Sometimes (actually fairly often) we found cracks in the engine cases themselves and had to replace the engine. Sometimes we replaced the engine because it was "high timed."

We got so good at changing engines we could do it in an hour fairly consistently. Strangely, one of the simplest "fixes" for a jet engine problem is simply to get another one and then let the engine shop figure it out and fix it. They don't like that, by the way.

One of the author's special activities was rigging the throttles for all four engines. Another special project was doing the inspection on the auxiliary power unit (APU). This was the same APU found on many airliners of the same 1960's vintage, 727, DC9, 737.

One thing special about the C-141 ISO docks that all of us there liked, was the fact that we usually had lots of free time on our hands. The work had to be done, and done fast, as we only had two or three days to do it, but we worked at a feverish pace because of that. And because of this feverish pace we usually got done early, if there was not too much work, which, I must admit, was often the case. It was feast or famine type work. We worked at a feverish pace because when we were done, we were done. The rest of the day was break, or we would simply go home. In the military you basically get paid a salary, so there is no need to stay on a clock. This was the incentive program our management used to make sure we stayed ahead of the schedule. We had a motto on a patch, which consisted of a rat with golf clubs in one hand and a toolbox in the other; the C-141 dock rats. That kind of describes our work area there. We got our work done, and efficiently, but, we had plenty of time to socialize with each other and we did it on and off work. Us dock rats formed quite a bond with each other. The author got to be friends with his coworkers better than in any of the engine shops. We were a tight bunch and we protected the secret of our good working conditions to the last. We appreciated our situation and worked to protect it.

Having grown up in an all-white Pacific Northwest community, this was where the author's world-view got greatly expanded. Contact with African-Americans, a Jamaican or two, Hispanic persons, people from the East Coast, South, Midwest, people who had been to Europe, Asia, and all over. We had time to talk and we did to great extent. It was a great education. These were some of the best times of the author's life, at least work wise.

Another group of people that I had the opportunity to talk to a lot at the docks, were several reservists who worked at the United Airlines Oakland maintenance base. The Bay Area is only forty miles away from Travis, and my last roommate in California actually was an A & P mechanic for United at San Francisco. So, I learned a lot about what the work was like at United Airlines while in the Air Force.

After a couple of years being assigned to the docks, the schedule was increased such that we had the plane for a whole week. This was like heaven because now we really got a lot of time off, and we had the additional duty of doing our own post-inspection engine check out runs out on the ramp every Friday. And for me, that was almost pure fun in the sunny Northern California afternoons. I recall one of these days…

There you are. You're out working with the flight line guys now. Big green birds everywhere. Lots of noise out here; behind the hangar door it was a little muffled at least.

You head out to the aircraft that you are going to trim that day; you note that the morning air is very cool for California, but it is the usual very dry. You get to the plane, some of the other guys are already there. They had tried to start the engines a little while ago, but there was a problem with one of the instruments in the flight deck, so they had to shut down everything and call up the instrument guys.

Now everything is ready to go again and you climb up to the flight deck and greet your cohorts. Your buddy calls up the tower and asks for the humidity, pressure altitude, and temperature so that you can make your calculations for the performance targets for the day. You've got the Jet Cal analyzer hooked up to one of the engines for your true pt7 reading. Since it's such a cool, dry day, your max. EPR (power of engine) target is very high 1.95. You look at your buddies and say she's really going to be jumping up and down today when we get to the high power settings.

The guy in the right seat calls up the tower and asks permission to do a power engine run.

"Travis Ground this is MAC aircraft ### on spot ### requesting clearance to do a two engine max. power run."

You can only run two engines to the max. TRT (takeoff rated thrust) because it is feared that you will jump over your wheel chocks and go careening down the taxiway if you run all four to power.

"MAC aircraft ### you are cleared for a two engine max. power run on spot ###."

"Travis ground, MAC aircraft ###, roger that."

Everybody in the fight deck is just a little more serious all of a sudden. The guy at the flight engineer panel turns the momentary switch that starts the APU (auxiliary power unit, a little turbine engine) the first shot of the hydraulic accumulator doesn't seem to spool it up, you think to yourself "start clutch must be slipping," the second accumulator is added in and the APU sputters to life. The guy sitting flight engineer panel makes sure the APU EGT (exhaust gas temperature) doesn't shoot past 767 deg. C. You warm up the APU a while. The guy sitting at the panel flips switches to open pneumatic valves. The guy sitting left seat is going through his pre-run checks. He clicks his mike button to talk to the ground spotter outside:

"Clear number one."

"Number one is clear."

He pulls out the start button for number one engine. The guy sitting panel says "pressure drop" indicating a pneumatic duct pressure drop signifying that air is going through the starter motor of number one engine.

The ground spotter says "Rotation."

The engine goes through the start cycle and spools up to idle. You go through the other three engines in the same manner. You let them warm up a while. The guy in charge of this little operation sends someone outside to do a leak check. The guy goes out to each idling engine and looks for leaks, he makes sure not to get too close to the front or rear of the engines. As he climbs back in to the flight deck he says "no leaks."

The guy in the left seat advances the throttles and you hear the engines spool up to part power. The nose of the plane starts to noticeably jump around. You remember how impressed you were the first time you went on engine runs. You thought it strange that another dude just like yourself was allowed to control this great air beast, even if you were just sitting on the ground in one place. After all, what would happen if you jumped over the chocks and went flinging down the taxiway? What if the parking brake failed?

Just one of these engines could power a large ship, and here this guy had the power of four of them in his palm, able to whimsically power them up, and bring them back down. These are good old-fashioned jet engines. Very, very loud. I remember the first time I held the throttles in my hand, feeling the power, feeling how I controlled the great thunder they were making. Knowing that thunder could be heard miles away. It makes you feel very powerful indeed.

We spend much of the next hour doing checks on the engines. We do checks on the bleed valves, we check the rate of acceleration of each engine, we check anti-ice valves, generator and CSD (constant speed drive, for the generator) operation, oil pressure, etc., etc. Part way through the hour we do our engine trim.

We trim the engines so that they all produce the same thrust at a given throttle position. One of the engines is low at part power. Today I am the one that is going out to adjust the trim. A reservist new guy is going to come out with me and observe what I'm doing.

You're supposed to bring the engines back to idle to make your trim adjustments, and then spool back up to part power and check your instruments again and again until you get your trim right. But your aircraft is in a remote part of the ramp where no one is usually keeping tabs on you, so you and the fella's decide that you are going to make all of your trim adjustments at part power, of 1.5 EPR, which feels for all practical purposes like *full thrust* to the casual bystander.

My partner and I step outside the aircraft. It is very loud out here. Even with our "Mickey Mouse ears," it's deafening. The newer planes are not half as loud. We walk down the side of the fuselage so as not to get sucked in, so to speak. We cut over to number two engine. Now, it is not only *loud*, but you can feel the power of the engine, the concussion it makes, *cut right through to your heart*. You feel in inside your chest, it can be very unnerving at first, and some people just plain can't do it. My partner gets a worried look on his face.

As I stick my Allen wrench into the bottom of the fuel control of number two engine, I look at my worried looking partner and remember what I thought of this little procedure the first time I did it. This jet engine is running for all purposes full blast, and I'm kneeling right underneath it.

Four feet in front of me is a very unpleasant death.

Ten feet behind me is a very unpleasant hot death.

Wow, this is *really cool.* What a job. I could really get used to this.

My partner however, does not seem to see the joy in all this. Now he is starting to worry me with his obvious twitchiness. I am afraid he will freak out and do something he will regret. I make my trim adjustment on number two engine real quick, and instead of going over to number one, I motion to go back to the fuselage and in the plane. After my partner is back inside I go out and finish the trim. After the trim operation we run two engines at a time to maximum thrust for five minutes per set and then bring them back to idle, and let them cool down. When we started our runs we had at least 42,600 lbs. of fuel on board, and now an hour later we have about 33,000 lbs. of fuel on board our old aircraft.

***

The Reagan military of the eighties was a different military than in previous decades in that you were not a draftee, conscript, or peasant—you were a member. The all-volunteer military of the eighties was fairly well led, comparatively speaking. That is to say, well led on the philosophical level at least, the greater "mission." But, it was run not the best on a practical, business, day-by-day level. The day-to-day industrial processes of the military did not function by their inherent design alone. They only functioned well when "forced through" by the personnel. The people "made it happen" as they say. The processes always looked good laying there in the book, but in real life they did not always work as advertised. Sometimes we would quip; "we made it happen even in spite of the system."

Desert Storm went well because Almighty God was with us, and the soldiers, seamen, and airmen were all "on board." The people made it happen, the *system* did not. Like the ancient Greek thinking-soldiers, when Americans have a feeling of justified cause, it is not likely that Americans will be defeated.

But, after Desert Storm the Air Force realized "wow, we're really all fouled up." The amazing thing was that they actually admitted it, at least at that time. They embarked on a new campaign, Total Quality Management (TQM) to rectify the inefficiencies of the system. As a young sergeant, I went to these classes and thought wow! This is really it, the answer. We were introduced to the teachings of Deming and the Japanese industry practices that we had all seen nearly take over industry. It was clear that implementing these ideas was going to be a hurdle, as the system tends to usurp middle management authority. But then, not long after Desert Storm, I got out of the military and went home.

Life after the military

When I was a young man, I was raised up going to church. I went to a small private Christian school that was very "evangelical." But then, still a teenager, I turned to the dark side I guess you could say, and did my own thing. In my last couple of years in the military, I had gotten divorced. And, during this same period of time I had a religious epiphany. I saw real world examples of "so as you sow, so also shall you reap," i.e. "what comes around, goes around." This was very traumatic for me, and during this time, and even slightly before, I had come to a reconciliation with God. I became a very religious person. After I got out of the Air Force in November of 1991, I was unemployed for a while, intentionally at first, but then a lot longer than I wanted. I spent a lot of time reading religion and history, which fascinated me greatly. I read all day every day. I went to school a little. I acquired my FAA power plant license. I toyed with starting a business. I worked at a small shop that worked on boat motors, bikes, and cars, so as to learn the small business game. I eventually got a job at an air-freight warehouse just to be near the airport.

A few years after the Air Force I got a job at Tramco, a subsidiary of the BFGoodrich Aerospace corporation. It was the summer of 1994. Tramco was an FAA Part 145 Repair Station. A Repair Station is a facility that is certified by the FAA to perform maintenance to airliners. It is the same type of work that an airline would perform itself to keep its fleet up to shape, and keep it legal with the FAA; the "hangar" work. Most all airliner maintenance used to be performed by the airlines themselves. But the cost saving frenzy created by airline deregulation has gradually caused a shift to the work being performed by third party "Repair Stations." This was originally done as a cost saving measure to get around the high price of union labor. One of the first successful airlines to employ such measures was Southwest Airlines.

The airlines still do much of their own day-to-day "light" maintenance, but the heavy work is almost exclusively done at Repair Stations these days. What the military would call depot level maintenance, and airliners have "D" checks. Airliners also have "C" checks, "B" checks, and "A" checks. "A" is the lightest, and "D" is the heaviest amount of work. "A" checks are the most frequent and take just a few hours or less, "D" checks take a month or more. There are some other designations as well but you get the idea.

A Repair Station is like a shipyard in scope, but it is also like an automotive dealership in a lot of ways. The airplane is worked on to a Maintenance Manual, a Structural Repair Manual, and Wiring Diagram Manuals. You look up parts in an Illustrated Parts Catalog, which has exploded views of all the parts and lists of part numbers. Just like at a car dealership. You might work to a Service Bulletin (an after the fact engineering change), just like at a car dealership. There are warranties involved. Individual subassemblies, referred to as components, are refurbished to an Overhaul Manual.

Every U.S. airline has an FAA approved maintenance program. The airline has a written manual of its policies and procedures, its forms, and how to fill out its forms, relating to the maintenance of its aircraft. The maintenance manuals the airline uses, from whatever manufacturer, are approved by the FAA, and they have to use those approved manuals on their aircraft. The repair station also has a general policies and procedures manual (and even their own repair documentation forms and paperwork, but these are seldom used). When working on an airline customer's airplane, the repair station uses both sets of data. The individual airline's manuals and so forth come first, however, as they are FAA approved for their particular aircraft. So even though a mechanic is employed by a repair station, he or she still works by the airline's manual, documentation, and record keeping system. So in a small sense the repair station mechanic works for the customer airline directly, he just gets paid less, in many cases.

It is actually better for the flying public if an airline has its work done at a repair station instead of in-house. The repair station makes its money by fixing things, the more the better. The airline makes its money by flying airplanes, the more the better. You get my meaning I hope.

Much of the work comes in the form of inspections. In fact, most of the work in a heavy check such as a D check is a result of an inspection. With airplanes, we do not like to fly them until they break while flying. You can see the wisdom in that can't you? In reality though, some airlines push the envelope more than others in that regard. But, the FAA has standards as to how often a plane must be inspected, and the airlines must follow these requirements either willfully or begrudgingly.

A large scope, such as a D check, jet project consists of several phases: incoming functional tests, access, inspection, repair, reassembly, rig, functional test, and test flight, if necessary. This is one "job," or project. Within this project, are hundreds, or thousands, of smaller jobs broken up into what are called routine jobs, and non-routine jobs. The routine jobs are scheduled work like accessing, inspections, removal of parts at time intervals to overhaul them, etc. The non-routine jobs are discrepancies encountered during the check in the normal course of inspection. For example, a routine inspection of a certain area will generate non-routine discrepancies found in that area. Each and every one of these smaller jobs are documented on their individual paperwork. Each and every one must be "signed off" by a licensed mechanic. These sign-off's have legal weight to them like a doctor's signature. Often two sign-offs are required, one from a mechanic, and one from an inspector. When an airplane is out flying around in normal service, all its maintenance is documented in the aircraft's individual logbook.

When an inspection is performed on a jet it begins by removing a bunch of access panels, then taking parts off of, and out of, the jet. Then the parts are cleaned, inspected, measured and etc. The airplane structure itself is inspected using various means, including visual inspection, borescope, ultrasound, x-ray, and eddy current. We are looking for defects such as cracks, corrosion, worn parts, parts rubbing against each other that are not supposed to be, and delaminated composite structure.

After the inspection phase is completed, repairs begin. At this time we look up the bad parts, finding them in the parts catalog, identifying the correct part numbers for that individual jet, the effectivity, or "effective" part number. Then you submit a request to a purchaser to find the part somewhere in the world. Utilizing the parts catalog is a skill unto itself and takes quite a bit of practice to get good at. Imagine how many parts are on an airliner; there are quite a few.

Taking the parts off the jet and putting them back on is a lot of work by itself, not even counting the repairs that get done when the jet is "accessed." On a D check about half the time is spent taking the jet apart and then putting it back together. The other half is spent fixing all the broken things found during the inspection, which happened after the "accessing." If the jet is well maintained the repair phase is quite light. And of course, if the jet is not well maintained it can be quite a chore.

I got hired on in 1994, for eight dollars an hour. That was a shockingly low amount of money I thought, but I was tired of working for minimum wage and being unemployed. I worked in the Special Projects maintenance line. This was a group that basically did the odds and ends aircraft projects. I was in the flight control shop. These are the people that work on the wings and tail, or empennage. At first I ended up doing a lot of interior work, seats and carpet. But then I worked on more of both the wings and the tail, doing a wide array of activities, but mostly light work. I did get to work on several different airplanes, such as the Boeing 727, 737, 747, 757, 767, Airbus A320, McDonnell Douglas DC-10, and MD-80. And I worked on many different airlines planes, such as Alaska Airlines, America West Airlines, Ansett Australia, Air New Zealand, Casino Express, EG & G, Evergreen, Frontier, LTU, FedEx, UPS, Sun Country, Southwest, Western Pacific, corporate jets and millionaire jets, and others I can't even remember.

Even though Tramco was situated next to Boeing Everett in the Pacific Northwest, very few people at Tramco were from the northwest. They were a mix from all over the world, aviation being their only commonality. People from just about everywhere including Africa, the Middle East, Asia, a couple from Europe, but mostly a bunch from the Midwestern United States. Any of these groups would find there way into management.

In addition to what most perceived as low pay, Tramco demanded much more than most of us were used to, as far as work produced. The place drove a mechanic hard. Even though working at Tramco was physically easier than loading trucks at the warehouse, the company could have made use of the banner "work will set you free" like the concentration camp at Auschwitz. They did not take to slackers. It was kind of like a small business attitude, where time is money. Are you making me any money right now? This was the atmosphere. Many of the leads were trained to be harsh, oppressive, and almost never satisfied with what someone accomplished. They themselves had been driven hard as mechanics. The atmosphere was harsh by normal corporate standards, union company standards, even military standards. It was more akin to an auto shop on a much larger scale, or a very busy construction site. I must admit, I didn't like the place at all at first, and thought of it as my "school" job.

When the outside employment opportunities were good, the turnover rate was horrendous. "Don't let the door hit you in the ass" was the order of the day. Sometimes they would give you a raise to keep you if they really liked you, but mostly it was a no quarter attitude. But, they would always take you back as well, if you were ready to work.

The company actually started out as a small business called Tramco, a company run by a hard driving entrepreneur; a cigar smoking former Boeing engineer and named Ron Crocket who was all about making money. He was not known for his compassion, and thought of aircraft mechanics as a bunch of "berry pickers," who were "dime a dozen." He was said to have fired a mechanic on the spot for complaining about his wages, paying the mechanic's remaining wages owed that week from money in his pocket. In the eighties most mechanics started at six dollars an hour. But, he was also known to give raises to mechanics who really worked their tails off and "made him some money." He created a culture that was all about performance, and survival of the fittest. The subjective performance evaluation was a key tool in promoting this culture. They used a system called the "stacking system." This system would see how mechanics "stacked-up" against each other. It would rate mechanics from the best to the worst, like a sports mentality. This is how raises would be apportioned.

The culture was competitive and aggressive, like a volunteer fire department. A mechanic who punched his lead in the face for getting a bad evaluation was not unheard of! But that same mechanic might not even be fired, if they made up!

When Tramco was taken over by the multi-billion dollar manufacturing corporation BFGoodich Aerospace, the company retained its Tramco culture and management, and even the Tramco name for a while. After some of the first big airlines went bankrupt, Tramco benefited from some of the people from the maintenance management of these. So, there was some expert and effective management from the technical standpoint, to get things going.

This was to be a period when a lot of creative energy was going to be applied to aircraft maintenance. On the one hand you had the scrappy raw gusto, with a decidedly small-business attitude, of the early Tramco period. On the other hand, you had the financial resources of a large corporation that saw a business opportunity in aircraft maintenance. They had built one of the best hangar facilities in the country to make their business a reality. The mergence of these two forces was destined to make some sparks fly. This was to be a serious effort.

There was to be a large influx of people at this time, mostly from the military. Most veteran mechanics had always been part of a larger organization where maintenance was merely a sub-part. Here the mechanic *was* the main event. The maintenance was center stage and we all had the sense of it. The working needs of the mechanic would be number one for a while. That one mechanic in twenty that was the technically proficient one in the military, could here surge forward unbound.

The taskmasters were hard, but the funny thing was, however, that when you built up to the Tramco rate and quality of work, it made you feel good. Like a pro athlete or something, you knew you had accomplished something. The work you accomplished had definite value to the organization, and you had the sense of that. But you were one who worked for cheap.

Every mechanic was supposed to have a minimum tool investment of $500 or so, but the average was $1000 or more, and went up to several thousand over the years. Some of the mechanics, however, even in the early days, had tool and box investments of 25k and more. And this represented people that made less than $15/hr. They were the serious ones. It is strange to think that this was simply to more effectively make money for their employer. But to them it was a matter of pride and professionalism. Having a tool for any job was very cool, plus it had the extra benefit of lessening the number of trips to the tool room. And it might even effect the quality of work you could do, which was also a matter of pride.

The big shiny toolbox filled with goodies was very prestigious to some. It was a bit like having the nice car in the parking lot. The toolbox was the mechanics own personal space, his desk, his office, his business. Most in the early Tramco stage were fairly poor, and a single stack Craftsman box was the norm. The taco cart was the box you worked up to. Some people would have more than one box. You wore the wheels out pushing them around, so hot rod wheels were a good upgrade. The taco carts would crack in the center, so invariably would get a riveted scab patch "Tramco mod" repair to reinforce this area. Would it have flush rivets, or protruding head? Many people would make add-ons out of sheet metal for their boxes, often coming in on their own time to do so. The high-end crowd had Mac and Snap-on boxes and tools, and you can sink quite a lot of money into these collections.

The buying and selling of tools was quite a market business at Tramco. Beyond the normal wrenches, sockets, and pliers, there are a lot of other expensive tools for working on aircraft. Boeing surplus flooded this market with plenty of fodder for trade.

In contrast to most of the rest of the heavy aircraft maintenance world, there was a marked emphasis on self-reliance and resourcefulness. You learn by doing was the thought of the time. It was understood that you were going to mess a few things up if you were busy. The mechanic was responsible for almost everything in his workday. The mechanic was supposed to handle problems himself first if at all possible. There were not too many "take your problem there" departments. The mechanic was the problem-solving department. And, could you hurry up solving that problem? I have another for you.

To some coming in from the outside the place seemed unrefined, chaotic, like an accident waiting to happen. There were a lot of new and inexperienced people. Many were used to a more refined and corporate environment. But, being skilled at your job, knowledgeable, and generally having technical smarts and ability were very prestigious and respected here. Tramco was very serious about aircraft maintenance, but at the very same time it was a rough environment.

You showed up for work. Maybe there was a crew huddle. You get your job card. You have to evaluate what must be accomplished. You get your references. You always start your job with the correct references, cardinal rule #1. You get the correct reference microfilm cartridge and attempt to load it into the reader machine. Sometimes this could be quite an accomplishment in itself, as the tapes and readers got used so much they fell into disrepair often. After you get the right manual, which is anywhere from a few pages to fifty, then you likely have to look up the parts. Finding parts in the IPC wasn't always a piece of cake, especially for a new guy. Then you go to the tool room, which also has some parts, and get what you need. There is another parts "store" which is an outlet for the warehouse, this is where you pick up parts that had to be ordered. Then you go do the job. Maybe you have to make a tool from sheet metal in the back-shop. Whatever it takes. Maybe you have to go to another building and get that one certain tool they have over there. You don't wait for it to be routed to you, you go and get it. Make it happen is the order of the day. Figure it out and make it happen. Use whatever resources that are around you. Don't be a whiner. And then, you have to sell your job to the QC inspector. Now you have to add tap dancing to your list of daily chores. Hurry up!

The leads you are working for are constantly making light of large and complex tasks. Everything is an "easy deal." "Easy deal, you should have it done in a couple of hours." "This is what I expect." "That only takes an hour." "Aren't you done yet?" "Mooooo…."

Today you have a structural repair job that is detailed and involves many small parts and precision drilled holes, utilizing sketchy SRM tech data. Tomorrow you are putting up three or four big heavy transmissions, and a slat actuator. Next day you are doing a careful rigging of flight control components and surfaces. Day after that you are working with a crane and crew of several to put up a large horizontal stabilizer or rudder that has to be located very precisely even though it is very large. Day after that you are farmed out to interior shop and you are installing half a plane of seats. Next day you are using your almost forgotten algebra to balance an aileron tab, and etc, and etc. Relentless. When you run out of work, which happens fairly often at this pace, you can go home unpaid or stay around and tidy things up. Maybe all the crew would go to Kodiak Ron's and have a few beers.

But the system is designed with motion in mind and there is plenty of energy. It is not oriented to looking busy and filling up eight hours. This is in the early nineties, the people who have been with the company since the early eighties or before talk about how everyone these days are lazy, and how slow things go. Of course, the people who sit in the chair the most talk about how they "used to be" the most industrious. Perception I guess.

There was the heroic myth of the "Tramco Man." A caricature of a mechanic that would work for cheap, take a beating again and again, obtain much tooling to make his employer money, and work at his craft with an unmatched zeal.

Sometimes the lack of work periods of time would be quite extended. The work was cyclic up and down. Back in those days they would find you work helping out in the tool room, in the warehouse, or even sorting and putting back into inventory all the expensive aircraft fasteners and blind rivets that they had salvaged from the hangar floor and the bellies of the jets. The idea was to be doing something useful or go home. They would make every effort to keep you busy though.

The company was divided up into Maintenance Lines, each Line having one jet at a time usually, with its own supervision and management. Each one was known to behave as a company unto itself. I suppose the group that most summed up the early years of "Tramco" culture was Maintenance Line Five. This was a group of extremely proud and skilled fellows doing D checks to UPS 727 freighters. The Boeing 727 was the bread and butter of the early Tramco days. Between UPS, Federal Express, Northwest, and United, among others, Tramco did a lot of '27's. The '27 was touted as a "real" mechanics airplane. Those intimately familiar with them were very proud of "their" jet, almost like the F-16 guys I had come across in the Air Force. UPS 727's reminded me of working on C-141's of the Military Airlift Command, drab, dirty, and all business. On Line Five they performed a lot of major and extensive work on those old '27's. The wrenches, drills, and rivet guns would step lively in this environment.

The leads of this line struck terror into the hearts of the young mechanics of my Special Projects group, who were mostly young and new. The Line Five leads could be brutal in their demands, and how they made their demands, to any new mechanic coming their way. Working for them really wasn't much fun for a new guy, and some would just outright quit. Think of political correctness, now take that political correctness and nuke it. In fact, the local mechanic vernacular was often a tonal language gleaning most thought meanings from the use of the words "dude," and the "f" word, especially when being personable and speaking as comrades "from the heart." "What is that bad thing you are doing?" translates "Dude!, like F___???" But "duuude, like

F_______!" conveys greatness or impressiveness. "You are exasperating me" translates "f__, dude." Of course the subtleties of facial expressions and gestures help convey one's point. Most basic thoughts can be communicated in this fashion.

This was a testosterone world to the extreme. These were some of the people with the 25k toolboxes. They were the marines of airline maintenance work, and they were extremely proud of the service they performed for the United Parcel Service. But if you could make them happy, there was no question you were good. Later on when the company had a kinder and gentler era under the BFGoodrich Aerospace banner in the mid nineties, and Line Five had been disbanded, many from Line Five would have legendary status within the company.

To get into management one simply had to have been a go-getter. Drill so many holes, then take your turn. Not much was required of management skills, just job knowledge of the jet. Rough-neighborhood-like strong-arm force worked for some. Some corners of the company resembled the feudal manor, their petty lords ruling their little domains with an iron fist. There was a perception of a group of "good 'ole boys" in management, this group was likened to a Harley-riding, sportsmen's hunting club. That being said, however, there was in general more of an openness between labor and management than at large union companies. One did not have the sense that management was plotting evil schemes, while smiling at you all the while. They just wanted the work done, it was that simple, more open and direct. And probably just about every large enterprise is going to have a perception of a good 'ole boy network. People of a like mind will always band together for mutual protection and benefit.

True, some individual management members had their own agendas, but in general it was a simple environment, politically. The middle management were former mechanics, and not particularly sophisticated in general. The mechanics and middle management were cultural and class equals, and few pretended otherwise.

They did have a business sense, however, in the small-business sort of way. An auto, boat, or motorcycle dealership comes to mind. It was a world that was to-the-point.

The Goodrich culture was a stark contrast to any union shop, or even a lot of the military. The people, who mostly came from the military, were made to believe that this was the "real world." Most people had not worked for an airline, and so did not have anything to compare with. There was a fair contingent of former Boeing employees now and again, and so the people learned what it was like at Boeing. But Boeing is a manufacturer and the airlines are about maintenance, so it was usually seen as comparing apples and oranges. Tramco was not really the real world though, at least not the real airline world. It was something unique and special in the American work experience. Tramco ended up being as much a *movement,* or state of mind, as anything else. The Tramco workforce voted down several union organizing attempts in the nineties. Pay and working conditions improved radically nonetheless under Goodrich in the later nineties, and a lot of this improvement was a result of simple competition for labor.

Tramco was populated in large part by people from Ronald Reagan's military of the 1980's, the same people that brought you the results of Desert Storm, and I think that this had a lot to do with Tramco culture. There was a high incidence of conservative Republicans. Religious and family values were not looked down upon. Gun collectors, hunters, NRA members everywhere. This is fairly common in aviation, actually, but here even more so. There were even a few sporting concealed heat, even though it was just as much against the rules as anywhere else. Despite emotions and tempers often flaring on the hangar floor, no workplace shootings, however (sorry to disappoint the anti-gun lobby).

One of the most influential groups of people at Tramco was a large group of young A & P school graduates from the northern Midwestern United States. Having started at Tramco, they had not been immediately corrupted by a large union company. They did not know they were supposed to sandbag.

Most people had the airlines built up grandly in their minds in those early days of the nineties. They thought you had to be a rocket scientist and athlete combined to work at a major airline. So they naively put their all into their efforts to be an airline mechanic. They did not realize that people at the actual airlines did not work that way at all. They believed the propaganda that they had come across at A & P School, or somewhere, of what was expected at an airline. The airlines always had a few people who could really strut their stuff, but this turned out to be a minority in actuality.

When the first people started getting jobs at some of the major airlines, there was a definite clash of cultures. High-strung Tramconites emerged at a few carriers and really made some waves with their ability and go-get-um attitude. Soon the low esteem of the repair station started to reverse at many airlines.

All the events of my work life up to this point started me thinking about the various organizational styles of labor, i.e. the company with a union work force working for an hourly wage, and the non-union hourly wage earner company. And what impressed me the most was how these two environments affected the people involved, and our country as a whole. My Christianity greatly affected my views. I began to wonder if I was on a "mission from God." I came to some conclusions, and I wrote a letter[2] to the Goodrich aircraft maintenance division's new president, John Martin, and expressed some ideas. He wrote me back with a favorable reply, but that was about the end of it. He asked me if I would be willing to go over my ideas with some of his subordinates, but then my career took another turn, right at this time.

If I stayed with an airline career, I had always thought that I would like to work for United Airlines. I had always thought of them as the big time when I was younger. In the late nineties it was the largest airline in the world. United was supposed to be doing great at that time. Their reputation was great as well, and they were supposed to be some of the world's foremost experts on heavy jet maintenance. A majority of their maintenance assets were based in the San Francisco Bay Area. No one had to work too hard there. In the 1980s I heard first hand accounts from reservists that I worked with at Travis that a United mechanic might go to work and do nothing all week.

In the wake of Desert Storm, in the early nineties, all the airlines were doing poorly. It was an aviation great depression. United and most of the other big carriers were losing money. In an attempt to keep their jobs, the United Airlines union employees bought a majority of stock, and it was touted as an employee owned airline. This supposedly really turned things around and made the place hum.

---

[2] The letter to John Martin is in Appendix A, p. 174.

I had applied with United several times in the mid-nineties after I got my A & P tickets, and then my FCC license.

At Goodrich I was really starting to fall into my own by 1997. I had been there just about three years and I was starting to do well. I was getting along better with the people around me, and I could have stayed there. But then United came knocking and I thought it was the opportunity of a lifetime, even though the starting pay was just a dollar more than I was making by this time. I left Goodrich in the middle of June, a few days shy of three years after I had started there in 1994. I was headed for the United wide body maintenance and overhaul facility in Oakland, California, United Oakland MOC.

When I arrived at United in Oakland, I was almost immediately dismayed. Within a day or two, still in orientation class I believe, I began to hear rumors about the corporate management wanting to farm out the heavy maintenance checks. And I could certainly see why. It was the most lethargic operation I had ever seen since I was in the Air Force. And in fact, it seemed a lot like the Travis AFB that I had been at just forty miles away. In fact, I think the Air Force was better. I saw a lot of familiar faces from Travis.

I soon learned that the medical benefits were not really any better than the place that I had just left, in fact, they seemed to me to be not as good. The pay was only slightly better, and the Bay Area at that time was a more expensive place to live, which left me at a net deficit. The flight benefits were only of nominal value, as the load factor was high at the time. It became clear right away that the flight benefit was not going to do me much good anytime soon.

There was a kind of haughtiness among some of the people. On the one hand they would bad mouth the place and admit that United had problems, but on the other hand they thought that United was as big and as fixed of an object as the government or military. In orientation class I made a reference to an airplane's tail number, the standard "license plate" of any plane. I was quickly informed that at United we referred to "nose numbers," not tail numbers. United needed different internal identity numbers than the rest of the aviation world for some reason. One got the impression that the FAA was not to interfere in United maintenance affairs, nor were manufacturers like Boeing. In the United Airlines history film presentation given in orientation, there was not a single mention of the fact that Boeing started United Airlines. Later on I would learn that some people would snidely refer to practices at United as the "United Way."

The hangar facility was mediocre at best. It was definitely less impressive than the Goodrich facilities in every way. The place was drab and dingy, the back shop equipment for working on the sheet metal jobs and so forth was old and unimpressive. Their tooling was unimpressive. Their ground equipment was also mediocre. I was getting under-whelmed in a hurry.

The attitude of the workforce was also drab. They were polite; this was a politically correct environment, no swearing for the most part, no bigotry, no sexual harassment. But beyond that, the place left a bit to be desired. And this wasn't really a genuine political correctness, but more of a decreed one. It was the nanny corporation, holding the stick over the head. We were directly in the shadow of the starkly segregated Bay Area, and not far from Berkeley.

The culture was a bit childish and naive. Being brand new, I was a "probie," or probationary worker. This ended up being like some kind of silly club initiation period, nothing more than a reason to mess with you a little. People kept on saying "you'll be here along time, if you pass your probation," with a little smile. In reality no one was expected to work hard, not even probies. You just had to know the rules. For a probie the most important thing was to strictly observe break time, and don't call in sick your first six months. Oh, and no sleeping. This is a big one and they emphasized this—no sleeping. Must have been a problem in the past I guess. Beyond that, not much was expected.

They did have a few interesting people there, their reputation garnered the place some of the best people that could be had in the field. And they did seek to hire the "best people available." But, these people would not thrive too well here.

The management seemed old and archaic. The foreman, the supervisor, assigned all the work, and micromanaged the operation. The lead was there simply to get the references and chase the parts. It was one of those environments where things were broken up into unnecessarily small factions. This person's job was this, and that person's job was that. The workers then "worked," thus provided with their materials and resources. This did not seem to work all that efficiently. No one wanted to be "bothered," and the environment was not very business-like.

One day when I was sent to work for another work section, I seemed to irritate the man in charge by looking up my own maintenance manual reference for the job that he had given me. He had not known that I was new when he had given me my job, and when he found out it seemed to bother him and he sent me back to my shop.

When I told some people that I wished to make some suggestions to my foreman about the sheet metal back shop, I was told that I should not do so, and that I should fear him. He was just there to get me I was informed. I needed to lay low, especially me being a "probie" and all. On the one hand I had been told in orientation that I should be a go-getter, but on the other hand I was told by my co-workers that I should not. This place was full of double speak and double standards from the word go. When I confided in a veteran employee that I was not too happy at United, at least at the Oakland base, he admitted that I was not alone, and that that was the norm.

Here I was at the biggest airline in the world, which was making money hand over fist at that time, and everyone was unhappy and all was chaos. And this was the supposedly *completely turned around* United, employee owned and all that.

They were very happy with my work quantity and quality, as they would have been with any of my Goodrich coworkers. A Goodrich person, at that time, blew them away. Our skill level, for time on the job, was far above theirs. Our knowledge of maintenance processes the same. And our quantity of work, produced in a day, was far above theirs also, without even trying hard. I remember a person there who was supposed to have been a great new worker, and she did have the personality of a get-it-done worker. Her job description was sheet metal. She had been there six months. She had not yet driven a blind rivet at work. I, a new guy, taught her how. At Goodrich, a person who had been there six months who had a job description of sheet metal would have done all kinds of things by that time, and would have shot hundreds, or more, of blind and solid rivets.

Well, after being at Goodrich and seeing how things went on Southwest, I thought I had made a big mistake. I thought I had jumped onto a sinking ship. So, even though my dream had been to work at United, I knew I would have a big trouble with joining the Union, morally. I decided that this whole situation must be a God thing and I decided to roll the dice. I wrote a letter[3] to the head person at that facility telling them how fouled up their corporate culture was and how it needed immediate, emergency attention, and I believe I gave them the letter I had written to Martin at Goodrich also.

I did not hear anything for a week or so, but then it hit the fan. You would have thought I had attacked a government official or something. I really freaked them out.

I got a message to go see my foreman in an office as I recall. We went to an office upstairs near the head office. They questioned me in a room full of managers and security guards. They asked me if I had actually been the one who wrote this and that in the letter. It was all very serious like in a government conspiracy movie; United seemed big on theatrics. All I could think was that I was sitting with a bunch of girls. Then they sent me home. They wanted me out of the building it seemed while they figured out what they were going to do with me.

The next day they sent me over to a flight surgeon in the San Francisco base, to head shrink me a little, and find out if I was crazy! I talked at length with the elderly doctor, and he was impressed with me, but told me that United was like the Queen Mary, and needed at least five

[3] The letter is featured in Appendix B, p. 183.

miles to turn around. He said that it was asking way too much to expect quick results.

In all of this no one ever said I was wrong, they just did not like being directly confronted with it.

Back at Oakland they were trying to figure out how to keep me out of the facility for a while. My foreman told me that he thought people would physically try to hurt me. Well, they ended up asking me to resign, which I accepted. Even though I had only been there exactly one month, they paid me a thousand dollars of severance pay, in addition to my regular pay. During this whole affair, small but significant miracles took place that aided me on my way; especially the weekend I wrote my letter to the boss.

Today, the work done at the Oakland facility is now done in South Korea. United used the events of September 11 to shed both the Oakland facility and another equally large facility in Indiana. United had well over 12,000 A&P mechanics in the 1990s, by 2007 the number was reduced to about 5,500 and dwindling. At the end of the first decade of the twenty-first century, United did no heavy maintenance in-house.

Chapter III

A Career with a New Purpose

When I came back to Goodrich three months later, it was to my same job as before, but now I was on a different maintenance line and all we did were heavy checks on Southwest Airlines 737's. I received the same pay as before I left, with the raise that I would have gotten if I had never left. I now had a different view of Goodrich; I realized what a unique place it was. I had a much more positive attitude about being there. And the company leader at the time, John Martin, was almost universally liked and appreciated, even though we did not always show it very much. Martin went to great lengths to change the negative attitudes at Goodrich. He was the Charlemagne that brought Goodrich out of its Dark Ages.

After several attempts to reform the much-disliked subjective evaluation process, Martin abolished the evaluation entirely, and went with a straight semi-annual wage increase. The top of the pay scales increased greatly. He made certain his staff understood that they would have to be more amicable to the worker population in the future, and would take action to this effect. "Everybody needs to get along" was his mantra. He sponsored several surveys, classes and events designed around positive attitude improvement. He was benevolent and sincere. But he seemed an effective businessman as well. Some would later refer to him as the "happy spender," but the fact remains that this period was the height of Goodrich aircraft maintenance in every way.

During this time, the parent BFGoodrich Aerospace Corporation decided to change its name to simply, Goodrich. Our division became Goodrich Aviation Technical Services, or Goodrich ATS.

This was the heyday, at least from the worker point of view. At that time it was not uncommon for people to leave Goodrich ATS, and then want to return. Very often the grass was not greener at this or that airline, or other maintenance facility. The period of the 1990's at Goodrich was characterized by near constant change and forward moving energy. The second half of the nineties were very dynamic. In 1998-9, the division was the largest it had ever been, the parking lots were over full. Martin would report that we had made at least a little money, and there were plenty of jets in the hangars.

During the same late nineties period, the governmental forces seemed to be determined to slap Goodrich ATS around. The business had attracted attention from everybody. The FAA came on strong, and we were judged with the strictest standard. The FAA couldn't find that much to complain about though. That just wouldn't do for some, and someone inside the FAA leaked information to the local papers that Goodrich was doing "disaster" quality aircraft maintenance. This had to have been purely politically motivated somehow, as nothing could have been farther from the truth. The official FAA position denounced the newspaper article source as unfounded, which it in fact was. Then there was the state industrial safety organization, and just about everybody else it seemed for a while. A full Bill Clinton era government assault, perpetrated by organizations that were curiously in Democrat control. They seemed very interested in this vibrant non-union company, which was situated mere yards from the state's, as well as one of the nation's, largest union companies. But each time we just said, "how high," and "thank you, may I please have another." Kind of the way things were at Goodrich over the years, really.

Yes, I had a different view of Goodrich, but the rest of the maintenance part of the airline world seemed completely headed for oblivion in my eyes. It did not seem to me that any of the large airlines were a great career choice at that time. I had seen the trend was toward outsourcing the heavy maintenance in trade publications, and I had now seen the reasons why in real life. And, I thought for real that I was on a mission from God. I used to say to my coworkers in the following years that I was on a "mission from Gad," like the characters would say in the movie *The Blues Brothers.* This was meant by me both humorous and serious, and was taken that way by people. You could get away talking like that at Goodrich, it was not so politically correct of a place. I thought my mission was to be a catalyst for a cultural evolution of the American corporate work experience. This evolution would be practical in nature, both work wise, and financially, but its result would be a return to more traditional Christian values. I wrote another letter[4] to Martin, and my letters to management eventually became known as the "Parker

[4] The second letter to the Goodrich aircraft maintenance division president is in Appendix C, p. 186.

Manifesto" by my coworkers in the same humorous, but serious, way.

By this time in my life I had grown quite comfortable in the Pacific Northwest, and I did not want to move far from home if I could help it. I kind of limited my potential career choices to Seattle, or somewhere close to Seattle by air. As luck would have it, I had another airline interview not long after coming back to Goodrich. I had applied at two different airlines in the last year, United and Alaska Airlines. Now I had an interview offer at Alaska. In the eighties, Alaska had seemed like the perfect career destination for an aircraft mechanic who was from the Northwest. Alaska has always offered a great product in their "flying transportation service," and a few of my Air Force buddies thought Alaska was the cream of the crop. Some people from my base had gotten jobs with Alaska around 1990. So I had had some feedback. Plus, by this time a few Goodrich guys had gotten jobs at Alaska. Alaska mechanics had been represented by the IAM in the past, but now they had a new union, the Aircraft Mechanics Fraternal Association (AMFA). A lot of airline mechanics had wanted a union that was comprised of mechanics only, and none of the peripheral people normally represented by the Machinists. The pilots and flight attendants both have unions like this, some of us thought this new AMFA might be a good thing for the future. Would they be?

Alaska was known to have some issues in their maintenance department. They had an attitude of attending to how things looked, not how they were, sometimes described as "smoke and mirrors." And, surprisingly, they were cheap when it came to replacing expensive parts and so forth, and this was seen at Goodrich also, when we worked on their jets. But, let there be a burned out light bulb in the passenger cabin or something like that and they were all over it. Their maintenance people at the Seattle maintenance base had a mediocre view of their employer at best, in many cases. The airline was kind of "all show and no go." Well, I thought maybe I could go down there and help them out with my new found radicalism. I really did not want a repeat of the United series of events, so I decided to make my pitch right in the interview. I had had a couple of interviews with Alaska before, for warehouse and ramp service type stuff. These were jobs that started at like seven dollars an hour, but you would think that you were applying for a vice president seat with Alaska. For these entry-level jobs the interview process was quite dramatic and haughty. And on top of that, I was never successful at getting one of these jobs.

But this time I knew I had the job, if only I would say the right things and play along, Goodrich people had made names for themselves wherever they went and the way was paved. But, like I said, I did not want to go to Alaska only to quit, or be driven out a short time later. So I interviewed with them and all was going well, I felt, I knew, that I had the job for sure. But, in an act—which has sort of become my trademark, I insinuated as politely as I could that Alaska had some issues in their maintenance culture and that I would like to come and be kind of an activist for change in their organization. They looked at me with dismay. I wish I had a picture of the expressions on their faces at that moment, it was priceless. I had dared to suggest that they were not perfect. Well, I guess I was not going to get that job, and that was o.k. A few days later they called me to tell me I didn't get the job, all I said was "o.k." It is interesting to note that about a year later Alaska crashed a jet for some of the stupidest reasons imaginable, mostly for hubris. I am not saying that hiring me would have changed that outcome, but what if?

By contrast, I have had a lot of contact with another of the nation's narrow body carriers, Southwest Airlines, who haven't had any fatal crashes. And even though their flying service is known as a bit of a Texas cattle drive, they take excellent care of their airplanes. They don't do much of it themselves, but they pay to have it done well by third party providers. Their attitude in the past has been, "if its bad, fix it, here's the money." This philosophy was implemented best by their original founder, but still continues. At least in the nineties, their corporate culture was almost dead opposite of Alaska, and Southwest employees enjoyed working for their employer.

Well, it looked like I was back at Goodrich to stay for a while, so I hunkered down and joined the team. After a while I was promoted to Master Mechanic, at the behest of some of my coworkers (thanks Lo). At first our maintenance line was doing light, one-week checks, but then before too long we were doing the month long heavy D checks. A month long cycle of work, over and over. You really learn what makes up the plane doing D checks; you rip almost everything apart, and put it back together and test it. The general phases of the heavy D check are incoming functional test, disassembly of almost the entire jet, inspection, repair, reassembly, adjustment, out going functional test, test flight, post test flight adjustments, and wipe and shine until the customer takes the jet and gives you a new one to do. Whether it be true or not, a mechanic working 737's at Goodrich in the first few years of the new millennium sometimes felt as if he were on the champion team, the best there was. Lets take a look at a 737 classic D-check, circa 2000…

So here it is, a new project, a dirty jet with a worn paint job. The jet arrives and we spring into action. As flight control shop, we go out and perform some incoming functional tests to the wings and tail. We have to do them now while fuel is still on board. The fuel is a coolant for the hydraulic system and we cannot run hydraulics for long without fuel in the tanks. At the same time we call for the fuel truck to come and top off the tanks. We need the fuel tanks full so we can see if there are any small fuel leaks anywhere.

We put a dial indicator on the rudder, or an aileron, with hydraulics on to hold the surface firmly in place. Then we push on the rudder with a scale to see if it moves a tiny bit. By doing this we can see if it has worn hinge bearings.

We put a torque wrench on the captain's control wheel to measure the breakout force necessary to separate its motion from the first officer's wheel motion. It has a mechanism that allows it to do this.

At about the same time other mechanics are opening panels on the lower surface of the leading edge of the wings so we can see if there are any fuel leaks on the front spar of the wings.

The first few days the jet is in check are a sight to behold. The jet is disassembled like it is in a pit stop at a *NASCAR* race. After many years of practice, the process is a certain…. poetry in motion. It is rush with skill. Simply rushing would not get it. Rushing without skill results in much damage. This was pure elimination of wasted motion (the goal of Lean). In the years after the millennium, we could do twice as much with half the people as in the nineties. In the nineties, a flight control crew might be ten or twelve people, after 9/11 you were lucky to have five. We learned to operate with a fluidity of motion that only comes with a personal responsibility for one's actions, an ownership. We each had our jobs, but we could often be found helping the other shops to get the job done as necessary. This was not really a result of Lean, but personal pride in one's work, a rare thing in America these days. A rigid process cannot give you this kind of efficiency in a changing environment, only a think-on-your-feet attitude coupled with a willingness to make it work. At one time in Goodrich you were given latitude on how to get the job done, as long as you were producing.

Perhaps a mechanism towards this end was the Goodrich Master Mechanic, a job position created in the mid nineties. The master mechanic was in the old idea of master, journeyman, and apprentice. And indeed Goodrich had all three of these positions. The old idea of the master, journeyman, and apprentice of course does not have them working for a major corporation. A master of a trade normally owned the business, and in turn employed journeymen and apprentices. So, this is at best an adaptation of the old paradigm. But it is one that seemed to work, at least to some extent.

It is interesting to note that many of the Goodrich master mechanics were in Reagan's military of the 1980's, where many were introduced to the ideals of Lean. This had the effect of producing a workforce that was somewhat accustomed to looking at things from the "out of the box" perspective. Innovative thought was the Goodrich way, unfortunately steadfast follow-through was not.

The master mechanic was supposed to represent a mechanic of above average ability, and the job description included leadership roles. The master mechanic was supposed to train the apprentices. The master mechanic would assume leadership roles on a temporary basis, as required, but the master mechanic was always a "working" job position. The master mechanic was supposed to be a technical expert. When the job was originally created, a mechanic was nominated for the position, either by popular opinion or at the bequest of his lead. The candidate would then sit before a board and be examined for merit. It is interesting to note that all other job positions would involve an interview with one individual, from the apprentice to the very top, except for the master mechanic who sat before a board. This did not completely eliminate people who were not worthy of the position, but in general the master mechanic represented the top percentage of the mechanic force. And another dynamic seemed to come into play. Normally when a person moves into management from a working position, they gradually loose touch with the working environment, and focus on the political aspects of their organization, for further advancement. This is great for them, but does not best serve the greater needs of accomplishing the work tasks. In many cases they did not want to work anymore anyway and that was there prime motivation for advancement. Of course this generally vexes the working populace. But with the master mechanic having a foot in both worlds, this was minimized to some extent. The master mechanic stayed well in touch with the nuances of the actual work, but was also involved with sequence of work decisions and so forth. The master mechanic, with a few coworkers, kind of assumed the role of a subcontractor working on individual projects. A supervisor would be in charge of an airplane or two in the entirety of the work accomplished. A lead, say a flight control lead, would be responsible for the work done on the wings and tail. And then a master mechanic would alternate running individual projects on the wings or tail. Normally in the flight control shop they had two master mechanics, one for the wings and one for the tail. Today the master mechanic is in charge of a few people removing the flaps, and is himself working to remove the flaps also. Tomorrow the master mechanic might be removing aileron parts on his own. But even then, an apprentice or journeyman might come to the master mechanic for help, instead of the lead, and etc.

Back to the D check—usually on day two we would remove the wing flaps. Those are the large things on the back of the wing that you see moving up and down on your flight during takeoff and landing. They help the plane lift better at slow airspeeds, together with the leading edge slats, by changing the shape of the wing, both in camber and area. There is more moving parts on these things than in the engine. Four flaps, each with a separate fore-flap and aft-flap, and linkages and mechanisms for extending and retracting these. Lots of shafts, transmissions, jackscrews, and moveable fairings we call canoe fairings. We call them canoe fairings because they look like canoes. The system also has position transmitters, a hydraulic drive unit, and a complex unit that controls and positions the flaps through the flap handle position in the flight deck. The flaps are heavy and beefy as airplane parts go, more machine like than the other flimsy sheet metal parts. They take a lot of air-load.

The flaps come off fairly quick these days. It used to take a couple of days and several people to get them off, but now we have it down to two or three people and one day to get all four of them off. We used to unbolt them to the point where they were ready to come off, and then place large worktables under the wings. We would then get about fifteen people to physically hold the flap while a couple of mechanics took out the last few bolts. Then amid much consternation and shouting, like pathetic pleas of "trailing edge up!" we would somehow pick it off the wing and get it on some sawhorses without dropping it. Every time it seemed as if we would trip and drop it. Your hands or arms would be on the verge of giving out by the time it was down.

Then one day somebody got the idea to use the proper lift sling and a crane. It took a long time to get used to doing it that way. The people who had grown up doing it the hard way did not like the crane. They would say, "see, we could have had it off by now if we just lifted it off." Of course these were leads and supervisors that were not doing the lifting. They used to do the lifting, and they thought everyone else should also.

But once we got good at the crane, what used to take many people, and lots of consternated effort to do, now took three people just minutes to do, and almost effortlessly. Not only easier, but no back injuries.

The ailerons are of course the panel-like flaps, way out on the back of the wing, that are used to steer the plane by tilting the plane laterally. They are one of the primary flight control systems, and are controlled by an array of mechanical components like cables on pulleys, linkages, power assisting hydraulic actuators, a trim actuator, and backup control mechanisms. They are also controlled by two hydraulic autopilot actuators, utilizing position sensors on the mechanical linkages, and the aircraft's inertial navigation system.

By day two someone is usually taking the entire aileron system apart. The ailerons themselves come off, taking a few hours apiece to remove if all goes right. You have to remove four hinge bolts, the balance bay hinge, the aileron balance tab rod end bolts, a position sensor rod, bonding jumpers, and the rod end bolt for the crank that actually moves the aileron. The ailerons themselves are very light, and therefore easily damaged, so great care must be taken. After many years on wing this is not always easy as the bolts are all stuck. A couple of guys then hold the aileron, while standing on ladders, rotate it vertically to clear the slot in the wing where they reside, and lift them clear of the jet.

All the linkage and some of the control cables come out for the check as well. There are two drums in the nose, under where the pilots sit, that come out. I did this job a thousand times I think. Cables connect these drums to the aileron linkage in the wheel well. All these linkage parts come out as well, including the hydraulic assist actuators for the ailerons.

Now is when you start opening the hydraulic system. The type of hydraulic fluid used by civilian airliners has some interesting properties. For instance, it makes a good paint stripper. The great air-beast has acid for blood. It is not supposed to hurt you, but it does afflict you with a burning sensation on the skin, so gloves are a good idea. If you get it in your eyes you will be running for the faucet quickly. And if you breathe a mist of this stuff, you'll want to clear the area very quickly.

The spoilers are square panels on top of the wing, towards the rear, that lift up to "spoil" the lift of the wing and cause that wing to dip. Each spoiler has it's own hydraulic actuator. Some of the spoilers work with the ailerons for steering, and some of the spoilers (called speedbrakes) are used with the wheel brakes for slowing the aircraft on the ground. The spoilers that are used for steering are tied in with the aileron system through linkages, and themselves are control by cables via a two-part gearbox called a spoiler mixer and ratio changer.

At about the same time the wing spoiler components come out, four actuators for the flight spoilers, two on each wing, and the spoiler control quadrant in the wheel well. If you are lucky, you will not have to change the spoiler mixer and ratio changer. The mixer and changer work as a proportional gearbox to move the spoiler linkage just the right amounts so that the spoiler panel angles are correct. The spoiler ratio changer is a real bear to change. Not fun.

When you remove all these parts, you have to clean them, tag them, and fill out the appropriate forms to have them routed to the overhaul shop to be individually overhauled or bench checked. Some of the parts would not be routed for overhaul and so had to be inspected on the spot, then put away for storage until ready for reinstallation at the tail end of the check.

During days two to three of the check, the fur is still flying all around. The jet is completely disassembled during this period. In flight control shop, all the wing flight control components, that are scheduled to be removed, are off by the end of day three.

During this same period the jet is jacked up by the systems shop, using four large jacks that look something like automotive jack stands, only the ones for a small 737 are six feet tall. After being jacked, large tail stands are brought in to make working on the empennage easier. If the jet is going to be greatly disassembled structurally, shoring is brought in and set up under the fuselage and wings. The shoring is very similar to what is used to support a ship that is out of water in a dry dock. The systems shop also removes the engines at this time.

On the tail the horizontal stabilizers, or tail planes, come off in this period as well. I bet you didn't think they could be unbolted did you? Also, if corrosion is detected in the mount boltholes, the vertical stabilizer comes of as well for some rework. All the removable panels, and the leading edges of the horizontals and vertical come off. The elevators are removed from the horizontal stabilizers after the horizontals are sitting on sawhorses. The rudder is removed. Then maybe the tail cone is removed. This provides access to remove the elevator operating components contained therein. If necessary, the elevator torque tubes and PCU assemblies are removed for inspection and repair.

In the tube, as we say of the fuselage, the fur is flying with just as serious a ferocity. By day two or three, all the seats have come out, the carpet, the bins, the sidewall panels, the galleys, and the lavatories. Then the floorboards all come up. The floorboards are light composite honeycomb panels. They flex under your feet, but they won't break. It's important that they don't break. The vital flight control cables run right underneath them in some places. If your foot went through the floor and stepped on a cable, you might inadvertently put the whole aircraft out of control.

Down in the bag pits the same is happening, the cargo pit walls and floors come out. Avionics guts the electronics bay, which has always been called the e-bay.

So, by the end of day three, if not a little sooner, the jet is completely accessed. It has been fueled and de-fueled. Incoming functional tests have been performed. It is on jacks, no interior, no floorboards, no engines, all access panels removed, no primary flight controls, no wing flaps, no horizontal stabilizers, maybe no vertical stabilizer, not many flight control mechanical components left installed, no black boxes, no flight deck gauges or pilot seats. It has been thoroughly ransacked, and looks quite odd without its tail. And all this in the good 'ole U.S.A.

Finally the pace slows as the inspection phase is entered. But now there is some pressure on the Quality Control inspectors to get the jet inspected so that we have some more work to do. Several inspectors perform a very detailed inspection of all the fuselage parts, the skin, stringers, frames, intercostals, floor beams, structure around the doors and windows, and all the "secondary" structure. Other inspectors are out on the wings, and up on the tail. When they are done with the aircraft itself, then they go and inspect the removed parts on the shelves in the back-shop. This physical inspection with their "eyes" takes a couple of days, usually.

Perhaps the non-destructive inspection (NDI) shop will have to look for defects, such as hidden cracks in a forging or something, using the eddy-current method or dye-penetrant.

Perhaps an x-ray of the wing spars or something would need to be performed and a special x-ray team would be brought in for this purpose. They use a near lethal amount of x-ray energy and so this must be done with all the people cleared away. So, it usually takes place at lunchtime, or in the middle of the night.

Once inspection has uncovered the defects, the repair phase begins. This might well require additional accessing. Very often in flight controls, on the wings, this meant that we would now remove some leading edge slats to access for work on the slats themselves, or the fixed leading edge of the wing that they are obscuring. The slats are moveable wing leading edge panels, the shape of the wing's leading edge. They work much like the flaps that are at the back of the wings. They increase the wing's size and camber to increase lift at low airspeeds. 737 slats are a pain to work on, -200's are the worst, the latter ones are a little better. They are difficult to adjust, and not the most fun to remove and reinstall either. But, a good mechanic who has removed them a few times can take one off in half a shift. A little longer to put one on, and potentially a whole lot longer to adjust them, if required. Fortunately, they do not necessarily need to be adjusted if they are simply removed and reinstalled for access. That is to say, simply removing and reinstalling (the same slat on as off) does not change their adjustment (yeah, that's right Sal!). If you have to adjust them, because you have a new one, or some other fit problems, the maintenance manual subject is longer than this whole book.

Maybe a fairing would need to come off, such as the number three or number six flap track fairing. Very likely at least one ground or flight spoiler would have to come off for underlying structure repair or bulb seal depressors removal and replacement. Sometimes nearly all the spoilers would be removed.

On the tail the horizontal stabilizer center section rear hinge points are inspected and refurbished. This requires shoring the fairly large and heavy center section with wood inside the empennage of the jet, as the hinges are the rear attach points of the stabilizer. If necessary, the elevator linkage is removed. This will require the removal of the tail cone of the airplane. It looks like a little hut when removed. To remove the tail cone the APU exhaust duct must first be removed. Systems shop just loves doing that. The elevator linkage is then removed, inspected and refurbished.

The wings are made as a box, with an upper skin, a lower skin, a front "spar," and a rear "spar." The spars are comprised of a middle web, an upper set of "L" angles called "chords," and lower chords. These spar parts together make a structure like an "I" beam. The upper and lower wing skins are the thickest and strongest skins on the jet by far. Attached to the front and rear of the wings are "secondary" structure. This secondary structure is less critical than the basic wing box. Also at the back of the wing is the "walking" beam that the landing gear is attached to and is the largest single chunk of metal on the airframe.

Wing repairs could run the gamut from very light, to extremely heavy. On the light end was the perennially worn "no step" panel aft attach point, which involved installing a new angle about a foot and a half long with a few fasteners. This panel is the one you see right outside the window on the aft end of the wing. On the extreme heavy end was an occasionally cracked wing rear spar upper spar chord, always cracked near the landing gear pivot point, which involved dismantling a large portion of the wing at the aft spar. This single project would take a small crew a couple of weeks to do.

In the lower main skin of the wing are oval fuel cell access panels. The flexing of the wing causes a small amount of wear at the ends of wing openings for these panels, and every jet has non-routine jobs to blend this "chaffing" out of these areas. The wing skins are a big deal and so this job is a big deal, with a specific and detailed process. It is not the most ergonomic job to perform, and so is more of a new guy job. It is not easy to hold your hands above your head for hours on end, blending with a die-grinder and then rotor-peening (a form of shot-peening) each cutout.

On the tail, they have their own set of repairs going. Almost every time, they are removing and replacing the bushings in the lugs where the removable horizontal and vertical stabilizers attach, especially the horizontals. This is a large, detailed, and precise job involving close tolerance machining on-the-jet, up on the tail stands. The people of our crew can do it in their sleep. Then there are the routine jobs like worn balance bay panel seals and the rubbing they cause inside the horizontal stab balance bays. Then the usual nicks, gouges, blown grease seals in the hinge bearings, and various other little things. Tending to minor dents in the removable leading edge skins is a normal activity. This is a precise process involving careful measurement, rework, and nondestructive testing for cracks. Sometimes a crack or sharp dent in the horizontal stab skin would necessitate a large skin section replacement. Happens every now and then. On metal jet parts, cracks and corrosion are the enemy. In composite structure, delamination of the composite layers/matrix is the enemy.

Do you know that lightning hits planes all the time? It hits one end and passes out the other end. Usually you see tiny little pits along the nose of the plane, they look like someone went along the side of the jet with a center punch, or poked it with a pen. Often there are several of these small pits on the nose, following a jagged line as the bolt of lighting danced along the side of the jet for that millisecond. On the tube these repairs are taken quite seriously, as this is pressurized skin, and that lightning heated the metal in those areas, like a welder. The area is drilled out and a small repair put in place. On the wing tips and stabilizer tips you will see the exit wounds.

Burned rivets and static wicks, not much to get exited about. Sometimes it blows a pretty good hole in Kevlar fairings.

One time a friend of mine who was inspecting the center tank of our jet at the time squeezed out of the center tank and told me he had found a bonding jumper burned in half. A bonding jumper is a wire used to ground various parts of the aircraft. He got a digital camera and went back in and took a picture (I was too fat to go in myself, they are small access holes). It was a large bonding jumper, about the size of a pencil, or larger, around. It was burned completely in half and the arc had pitted the adjacent sump structure with a pit the size of a pencil eraser. This arcing had occurred directly in the fuel in the bottom of the tank. This was a larger spark than is used to start the engines. But it had obviously not ignited the fuel. On the very same jet, I saw with my own eyes the charred burn marks left by electrical arcing on the fuel probes in the outer wing tanks. Out here would be just fuel vapors, but again, no blown up plane. This was not too long after the flight 800 incident, and it gave me pause. They reported the incident to Boeing engineering that evening, on second shift. The next morning, people from Boeing came and replaced the affected parts. I never heard anything about it ever again. Hmm.

The tube involves a large quantity of sheet metal work and repairs. The tube will be crawling with structures mechanics throughout the check. Often times whole skin sections would be replaced, especially the aft lower skins, under the aft cargo compartment. Around the door openings, especially at the corners, large and somewhat complex repairs would be installed. Some of them looked like the state of Texas, and earned the moniker "Texas

Patch," both for size and shape. Inside the tube a myriad of repairs are taking place, corrosion being the main culprit. The parts involved are frames, stringers, intercostals, floor beams, seat tracks, door thresholds, and etc. This involves blending out the corrosion with a die-grinder, and then treating the area with alodine and primer. If the corrosion is bad, or there are cracks, a double patch is employed or the part is simply replaced with a new one. In either case the rivets have to be drilled out, holes drilled, and new rivets or special fasteners reinstalled. Some of the parts are bought preformed, and other parts are made in the back shop to the specifications of the applicable drawing and Structural Repair Manual (SRM).

The better airframe mechanics get very skilled at making new parts with various angles and bends in them, using tools such as shear, brake, shrinker, spreader, and the English Wheel. Custom made tools are often employed also. These people are very creative, and handy with their hands.

Not all of the repairs are covered in the SRM, so the senior structures mechanics work constantly with engineering orders. With experience, they become so in tuned with their work that, at least sometimes, they essentially tell the engineers what needs to be accomplished in a given repair, and the engineers assume a role of verifying the suggested repair is correct, using their statics and dynamics studies. This relationship saves a lot of time and effort.

All this repair work generates a lot of noise: cutting, banging, riveting. With a giant hangar containing half a dozen jets in it all being worked on simultaneously, it literally sounds like a war zone. Rivets being driven here there and all over gives the hangar a battlefield sound effect. Sometimes almost continuous sporadic gunfire, and often it sounds just like the sporadic automatic gunfire like you would hear in a war movie. It is enough to make a veteran look over his shoulder. And it can really get on your nerves from time to time, but you learn to deal with it. That is what earplugs are for. But, especially after a lull, sometimes the noise is somewhat comforting in all the activity. It is the sound of energy and job security, a sign that things are moving.

In the repair phase and in the reassembly and testing phase of the check, a new dynamic comes into play: the quality control (QC) inspector. The inspector works for the company, same as the mechanic. Their job is two fold: they perform inspections on the aircraft to generate work cards that need to be performed, and they inspect and certify the work that is performed. Their job is to verify the work is completed per the Maintenance Manual, Overhaul Manual, and Structural Repair Manual specifications. They also verify that the correct parts and materials are received and used on the jobs. They do not perform any of the work. The idea is that they will be critical of the work, since they do not have to perform it themselves, but they are still completely liable for the work they sign off. So in a sense, they partner with a mechanic on a job to say that it is complete. The mechanic does the work and signs it off, and the inspector says yes, it is complete. The second set of eyes and signature, and this carries legal weight with it.

An inspector will "buy" items like the fabrication of a repair part, that it is made of the correct material whose origins can be verified, that the holes in the part are of the correct size and placement, and that the part is finished properly. When it comes to mechanical systems in the aircraft, an inspector will by adjustments, gaps, and torques, among other things.

Some items are classified as Required Inspection Items (RII). These are critical aircraft systems like the flight controls and engines. Maintenance Manual chapter 27 is flight control. Almost everything in here is RII. So that means that every time the manual references a nominal condition with a tolerance, that is a precise measurement with a (+/-) value, this condition must be verified by an inspector. Also, a torque of a bolt in a critical flight control, such as an aileron hinge bolt, must be witnessed by an inspector. This also applies to tests, and an inspector might need to see a voltage, or a value on the BITE computer screen (Built In Test Equipment). And of course, the inspector will look over an entire installation of a component.

So, putting the aircraft back together already takes longer than taking it apart, and, on top of that it takes even longer because of the need for a mechanic to chase down the QC continually for every little thing. Learning how to operate in this environment takes a combination of a great deal of finesse, will power, and cooperation. Much subjective opinion and bantering comes into play, as the situation is not always cut and dried to all involved. And the relationship of mechanic to inspector often is one akin to a used car salesman trying his best to sell a car to an unwilling consumer. The mechanic rationalizes; the inspector stoically argues the high standard of the maintenance manual, as if it were the Holy Scriptures. And on and on it goes. A curios thing one sees is that the most hack of a mechanic often becomes the hardest, most difficult to deal with inspector, and the best mechanic often becomes a reasonable inspector. That guy who was a hack of a mechanic, and connives his way into quality control, suddenly knows every jot and title of every maintenance document that ever was. And as an aside, that is kind of the story of airline aircraft maintenance in general, lots of bullshit. This is a human drama world unto itself. Those familiar with the Bible will recall the story of the Pharisees and Sadducees.

The flap control system would be adjusted within the first couple of weeks of the check. The control stand that is between the pilot and first officer would undergo an inspection the first week of the check and the flap handle with its attached cable system would be adjusted then, if the flap control unit had not come out. The flap control unit is at the other end of the flap body cables, in the wheel well. The flap control unit is a confluence of hydraulic valves, electric position switches, control cable input and cable output. If it has to come out it is quite a job to remove, reinstall, and adjust. The hydraulic valves control trailing edge flap position and have to be painstakingly adjusted, the switches have to be adjusted and the systems they relate too tested, and of course the control cables have to be rigged. The functional tests relating to the switches can take a week themselves.

After a couple of weeks, the wing repairs are done or almost done, and the flaps and ailerons that were sent out for overhaul start to come back. If all the flap tracks are now installed, and the three and six flap track fairings are on, it is now time to hang the flaps. Again we use a crane, and two expert mechanics, with one more mechanic on call for the actual moment of truth when the flap is hung. Sometimes we get a luxury crew of four, and if they are all experts, we get the flaps on roughly twice as fast. With two mechanics we get all four flaps on by the end of the day, with four we can get them on by lunch. That is because two are on the left wing and two on the right. If all four were on one wing, it would probably actually slow us down. And if a couple of the mechanics are novices, it will slow us down almost certainly, unless they stayed completely out of the way. At this time the flaps are not installed, however, just hanging on the jet by a thread. Installing the flaps is far more extensive than removing them. Installing them will take a crew of four a week, actually two on day shift, and two on night shift. Often one shift would take the inboard flaps mostly, and the other the outboards mostly. But they would mix it up as necessary to keep things rolling.

The wings are the most mechanically complex system on the plane; they have the most moving parts, more than the engine. The wing trailing edge flaps are just one of the wings mechanical systems. There are a myriad of clearances, adjustments, and spring forces to set.

On the wing outboard trailing edge flaps, the large pieces are the fore flap, mid flap, aft flap, and the flap track fairings that we call canoe fairings because they look sort of like canoes. The inboards on a Classic don't have canoes. The flaps are attached to a "carriage" which has roller bearings in it that ride along a track.

The flaps are driven by a system of drive shafts called torque tubes. These torque-tubes are connected by several transmissions of various types and sizes. In addition to the torque-tubes are eight transmissions driving ballscrews, commonly called jackscrews, that are attached to the flap carriages and move the flaps up and down the flap tracks on the wing. You can see many of these components when you are on a flight and sitting behind the wing. The torque-tube system is driven by a central hydraulic drive motor via a reduction gearbox. All these torque tubes and gearboxes have to be synced together, or “timed.” This is so the flaps will move properly. If you try to move them without having them timed, you can damage the flaps or the adjacent structure.

In the first years after the millennium, installing a set of flaps on a ’37 at Goodrich ATS was a beautiful thing to behold. The process had achieved a maturity: fluidity of motion, waste almost entirely absent, effortless speed, and excellent quality. Magnificent speed, and I would wager, unmatched quality. And it was achieved without overly stressing the workforce involved. I doubt it could have been done much better. Was this the result of Lean processes? It sounds like it would have been, but it was not. It was the result of the technicians involved taking ownership of the process and making it their own. Like it was their own small business.

In the mid 1990’s, just getting the flaps on and having them not break themselves when moving was about the best that could be expected. And that was with a crew twice as large, taking twice as long. The outboard flaps of a ’37 are particularly difficult to adjust for the novice. When the flaps are in the retracted position, you have what amounts to three different “spring” forces acting on each other in different directions, and you have to tune and adjust the flaps to get this correct preload in different directions, measured with spring scales and gaps between the flap parts. On top of this you have to set the Wing Chord Plane (WCP), which amounts to tuning the wings aerofoil shape to its optimum for airplane performance, and the correct trim characteristics. The trim characteristic is that the airplane must fly level on its longitudinal axis, or not tip to one side or the other, when the pilot’s hands are off the wheel, and hydraulics are off. The ’37 mechanics at Goodrich were very good at achieving this optimum condition.

The ailerons would go back together during this period also. In fact, on the wing the rigging order is generally flaps, ailerons, and spoilers. Leading edge slat work fits in where it can. During the first two weeks of the check the various component assemblies, generally referred to as off-units in the industry, get overhauled in the component overhaul shop. Then they start to trickle back to the receiving area at the aircraft one by one.

At the beginning of the check, the aileron system was completely removed from the aircraft with the exception of the aileron autopilot actuators. The fuselage aileron control system is under the level of the floor and forms a rectangle, with a component in each corner. The captain's and first officers control wheels are above the front two corners. There are two cable runs, of two cables each, for the aileron fuselage cables, one on the right and one on the left. They of course have their own individual mechanisms. The left side, or captain's side, has the captains control drum at the front of the cable run and the aileron centering mechanism at the rear. These are fairly complex parts that have various little do-hickeys all over them; position sensors, force transducers, and the aileron trim actuator. They attach to the cables via a quadrant at each end. On the right side we have the aileron transfer mechanism at the front, and a quadrant at the rear that is also attached to the flight spoiler control parts. At the front end, the captain's control drum and aileron transfer mechanism are also connected together via another set of flight control cables. At the rear end the aileron control system is interconnected by linkage. The two sides are designed that they can be controlled separately to some degree; if one side or the other became jammed. We actually test this feature during the functional phase of the operation.

The wing aileron system and the fuselage aileron system interface in the wheel well between the two rear "corners" of the fuselage aileron system. This is where the aileron power assist actuators, actually called power control units (PCU's), are located. Yes, they are really just assist actuators, the pilot can fly the airplane fairly well without them functioning if necessary. Nice feature of this "little" airliner. With large aircraft, you cannot take the name for anything for granted. Incidentally, every time they design a new plane, they often like to change the names of the equipment for no good reason. I think they do this to irritate and confuse the people that have to work on them. So on one plane an actuator is a PCU, and another it might be a PCA, or simply an "actuator," you just never know. You always have to familiarize yourself with your particular equipment.

Reassembly of the fuselage aileron system starts by installing the components at the four corners of the aforementioned rectangle. This is where that childhood playing with *Legos* really comes into play. You'll have to know how to use your tools here. A little tricky, but once you've done them a few times it doesn't seem so bad. Then comes the aileron "body" cable hook up and adjustment, or rig. First, you consult the maintenance manual rig tension chart, which not only gives you a rig tension for the particular system cables, but an adjusted tension that accounts for the ambient temperature. The fuselage is made of aluminum, and the cables are steel so the tension varies with temperature due to the different rates of expansion for the two metals. When you rig cables you typically have to insert a rig pin in the two opposite quadrants. One rig pin holds the front quadrant in place, and the same for the rear. Then, theoretically, you tension the two cables, using the turnbuckles, to an equal tension. To do this you install a special clamp on the two ends of the cables that holds them still, and turn the turnbuckle with a special tool. This is a precise bit of work and one that the average Goodrich expert flight control mechanic took great pride in doing. In reality the job takes a bit more finesse. In the '37 body aileron system, you have three separate sets of cables that work interrelated with each other. You want the rig pins to fall out of their holes without dragging, and equal tension on the cables, within a couple pounds. You have to adjust, pull the rig pins, cycle the system to work out the little friction error that exists due to the cables running across pulleys and quadrants. You then check it again and again until you get it dead nuts. In addition to the cable rigging, there is adjustment of the PCU linkage (two sets) involved also, which is in direct relationship to the cable rig. If you have to readjust one, you must do the other also. It is a bit like tuning a piano. After all this precise rigging, you will then have to "safety" the cables, or secure the cable turnbuckles so they cannot turn. This is done by aligning a slot in the threaded swaged-on cable ends and the turnbuckles themselves, and inserting a locking clip. The clip is a few inches long and sometimes won't go all the way in the slot due to some crud being in there. I hate it when that happens! If you can't get it to go after using all your tricks, you will have to break down your rig and clean the turnbuckle threads. Bummer! Don't even think about cutting the clip short so it will look like it's all the way in!

After you get the aileron body cable system right, next come the wing cables. First, step back and say a small prayer of thanksgiving that you are not rigging 727 wing aileron cables. The 727 has four ailerons, and the wings are much more involved to rig than a '37.

The 737 wing aileron cables rig a little different than the body cables. You put the rig pin in the center PCU linkage, which is the inboard end of each wing aileron cable run. Instead of putting a rig pin in the outer wing aileron cable quadrant, you actually measure the position of the aileron in relation to the wing. This allows for a very precise cable and aileron rig. One cable tighter makes the aileron move up slightly. The other cable tighter moves the aileron down slightly, and that is how you dial it in. The tolerance is that the aileron will be less than fifty thousands of an inch up or down from a fixed reference mark on the wing. Relatively easy, actually.

Once your ailerons are rigged, you can rig the flight spoiler cables. This is the trickiest cable system yet. There aren't the body cables to worry about, the flight spoilers take their control input from the aileron system at the back end in the wheel well, but there is a difficult to adjust, multi-cable system on the wings. The system is kind of opposite the ailerons in that way.

You have four cables and two quadrants on each wing, and the cables are physically very close to each other, or covered by equipment. This makes it difficult to install the turnbuckle clamps that hold the cable ends as you rotate the turnbuckles. Along with the difficult access, there is a lot of friction error in the system created by several pulleys, and the two quadrants. This arrangement requires that you will adjust and cycle many times to get the rig pins loose fitting in the two quadrants.

After 2001, the tail crew was usually only one or two people per shift. This person has to be a bit of an expert, and be able to accomplish quite a lot on their own. They had to take very definite ownership of their job, making it almost a business of their own. In return for this, they were pretty much left to their own devices.

On the tail, after the stabilizer attach bushing work has been completed, and the other repairs and mods, the elevator torque tube assemblies will go back on, if removed. This amounts to two large and heavy shafts, and a bunch of assorted linkage for the elevator PCU's. These are the same PCU's as the ailerons use, by the way, and they are assist actuators. Just like with the ailerons, the pilot can still control the elevators without hydraulics. After this equipment is installed, the four elevator cables, two on each side, can now be rigged. Access for these is pretty good, and rigging them is not very hard. After the elevator torque tubes are in, the aircraft tail cone can be reinstalled and this makes way to put the horizontal stabilizers back on. If the tail stands are used on this check, a hoist on the tail stand is used to put the stabs in position. Two people can easily install a horizontal stab. It bolts up to the plane with five very large bolts. Most of the time the elevator is already installed on the stab when the stab is installed to the plane.

If the vertical fin has been removed, reinstalling this is quite a bit more of a big deal than the horizontals. The tail guys from a couple of crews will likely band together for installing a vertical. A large crane is required, and only an expert crane operator and crew can pull this job off without a hitch. It is very much a job of great finesse. Again, the rudder will likely already be installed onto the vertical when it is put on the jet. The vertical fin bolts onto the jet with six large bolts. And of course the rudder cables will have to be rigged, and rudder PCU's adjusted. One main PCU and one standby PCU.

The tube reassembly goes through the last couple weeks of the check. First floor boards and side wall panels, then bins, galleys, and lavatories. The interior goes in very fast, huge visual progress. When it comes to putting the seats back in, they use a baggage conveyor parked up next to one of the service doors and put in the seats like as on an assembly line. The conveyor moves almost continuously all day, and a whole planeload of seats are installed in a very short period of time. And, the work is done in a harmonious fashion so as to not overly tax anyone; again, rush with skill. The flight deck goes back together at this time, but is a little more tedious. Only one or two mechanics are assigned to work on the flight deck. The flight deck is worked on by both upper deck structures mechanics, and avionics shop people. All the instruments and so forth usually come out of the flight deck so the underlying structure can be inspected, and then is reinstalled by avionics shop afterwards. Avionics puts all the black boxes back in the e-bay and gives them a check out.

Outside the jet, the systems shop reinstalls the landing gear if they removed them for some reason, and installs the engines. Installing an engine is really not all that difficult of a task. It does require expert handling and coordination to not cause any damage to the very expensive engines, but the engines are designed to be removed, and then reinstalled, fairly quickly. The engines are handled with a crane and winch system that bolts directly to the wing pylon, called a bootstrap system.

During the last two weeks of the check the flight control adjustments that require hydraulics on are done. The tests of the various flight control systems are also done during this late period of the check.

Once we have electrical power, we set the flap position transmitters. This is needed not only for the flap indication to work properly, but also so several other tests can be accomplished. Most notably of these are the landing gear warning tests. The system guys usually do these and it involves moving the flaps and landing gear and watching the bells and whistles in the flight deck. On some of the more complex functionals you might have to have a mechanic hold a switch or something down in the e-bay while the guy in the flight deck moves the handle for this or that, or pushes a button for something, and watches for this light, or listens for that horn. You really have to know what you are doing and be focused during tests like these to accomplish them in a timely fashion, and satisfy the inspectors that they passed correctly. In fact, multi-mechanic flight control adjustments and multi-mechanic functionals require a near perfect synchronicity among the participants to get them accomplished in a timely fashion. Inexperience, impatience, and lack of technical ability increase the time to complete these jobs almost exponentially, and with great frustration.

Complex aircraft adjustments and functional tests rely greatly on the right psyche in their efficient performance. A prideful attitude, i.e. trying to show that "you know what you are doing," and that you are "cool" can be a great detriment to results. A humble and friendly attitude is the start. You love your comrades, and you know they can do their jobs, that is the attitude the job leader projects. The job leader has to be technically adept at the job at hand, but if he/she in unsure about something, the job leader must be open about this and not try to "B.S." people.

Extraneous banter, joking, and communications are always the problem to be dealt with, but a nasty and impatient demeanor in addressing this is equally a project stopper. You are quite literally walking a tightrope here. The road to success is a very narrow one. A serious demeanor, but not too serious. Focused, yes. Angry and impatient, no. Flustered, no. Light humor is acceptable from confident and expert technicians. If you are a newbie, shut up and do as you are told, and try to keep up.

The aileron functionals are the first of the big wing flight control functional test groups. They take half a day to more than a day depending on the troubles encountered. We test for the amount of friction in the system with hydraulics off, and with hydraulics on. We check for freedom of movement throughout the full aileron travel. We check that the ailerons move up and down a specified number of inches. We check the function of the emergency equipment that allows one control column to still move if the other one becomes jammed. We check the aileron centering. That is we make sure the control wheel comes back to within a degree or two of center when released from a certain number of degrees right or left. We do these checks with both hydraulic systems on, then we isolate one out, then the other. And all these checks are done with measuring equipment; they are not just done to "how it feels." If the aileron centering is off a little bit, we might have to adjust the left side aileron cables, due to a slight peculiarity of the systems design. If we do that, then we have to readjust linkage on the aileron PCU's also. But this is what it takes to get it right and make for a great test flight.

Next comes the spoiler adjustments generally referred to as "pick-up and lay-down." Lay-down refers to how the spoilers sit when they are in the down position. They have to have a specific gap between themselves and the flaps, which they are directly over. Pick-up refers to when the spoilers start to move up when the control wheel is turned right or left. When the captain's control wheel is turned right or left, the ailerons of course move accordingly. But the spoilers start to move at about ten degrees of aileron travel also. This gives "extra" turning power to control the plane. We check the angles that the spoilers are supposed to come up to. This includes the time when the spoiler first starts to "pick up," or move, and the fully extended angles, which represent maximum turning and speedbrake extended angles.

We adjust the spoiler pick up first, by doing a procedure called "counting off" to the right or left. A mechanic in the flight deck has a protractor mounted on one of the control wheels. He then turns the wheel slowly and, using a radio, counts off degrees to a mechanic behind the spoiler on the wing. Five, six, seven, eight…and counts off until about fifteen degrees and then stops and holds for a moment. The guy on the wing is watching the spoiler while listening to the count. He is waiting to see the spoiler start to move when a specific degree value is reached. If it is picking up late or early, the wing mechanic will have the spoilers raised fully, and he will adjust the spoiler servo actuator linkage a little, and repeat the procedure until it is dialed in. After the pickup is set, the extended angle checks are accomplished. If they do not come up to par, there is nothing to adjust to compensate, and you will now be looking at changing a worn spoiler ratio changer and spoiler mixer…bad day.

The wing trailing edge devices functionals are the big focus of the wing adjustment and test phase, but the wing leading edge slats and kreuger flaps will be fit in where they can. The leading edge is another maintenance manual subject that is something like eighty pages long, but you can verify the slats are in adjustment using the beginning of the manual if you have not changed anything about the adjustment of the slats. They simply move out and in, stopping at an intermediate position at certain trailing edge flap handle positions. The trailing edge flaps control the leading edge devices. You check that they move freely, and there is no binding. You check that the gap between the wing and the slats is correct. If you have to adjust this it can be a real chore. The slats fit against the wing with a very small clearance, like they were an engine part or something. The tolerances are indicative of a machined part made of solid aluminum, going onto a solid wing. But the slat is made of sheet metal and composite honeycomb structure, and the wing leading edge is composite structure, and so both get a little twisted out of shape during the years of constant wing flapping. Both the wing and the slats are of course large. This can make adjusting them ten years from when they were first manufactured and adjusted a bit of a chore at times. The inside of the slats and the wing leading edge are painted with Teflon paint incase they rub a little while flexing in flight.

After the basic aileron and spoiler functionals are accomplished, we start checking the bells & whistles in the flight deck that are related to these systems. Among these are aileron roll force transducers adjustment, aileron autopilot authority and aileron force limiter checks. I must admit, I always enjoyed doing these, but you do have to be very focused when accomplishing these tasks to get them done in a timely fashion and make them pass o.k. They are done using the airplane's BITE computer, and some measuring tools. You are checking the health of the aileron autopilot actuators, and the adjustment of the sensors in the system, as well the general functionality of the system including the Flight Control Computers (FCC's) in the Digital Flight Control System (DFCS). If all goes well, and no adjustments are required, we can have it done by lunch easy.

The maintenance manual subject for adjustment/test of the elevator system is eighty pages long, and that is standard 8.5 by 11 inch pages. It is the longest adjustment/test manual subject of any of the flight controls. And it is tricky to adjust properly and have it perform well in flight. It will be tested in a test flight, and so its correct adjustment is of high importance. In fact, it is the number one flight test issue normally. Wing rolling is less common than miss-adjusted elevator tabs, causing pitching, whether up or down.

Next comes the rig of the rudder system, which is not too difficult for the tail people. The horizontal stabs get their functional check out, they are electrically operated and sometimes require an extensive functional.

After that we get into some of the extensive elevator functionals, like elevator feel-force and elevator autopilot authority. These involve hooking up a calibrated pressure device to the pitot-static system probes on the tail, to simulate airspeed. The elevator has a system that varies the pressure the pilot feels when he pulls back on the yoke, which depends on the airspeed. This simulates what a non-hydraulically assisted elevator would feel like. The faster the airplane is going, the more air pressure on the elevator, and therefore harder to move. The functionals for this system take a couple of days of very focused multi-mechanic interaction, using the airplane's BITE computer. For feel-force, you put on an airspeed with the test box, and then push or pull on the control yoke with a certain force using a spring scale, and see how far the elevator moves in relation to the tail cone neutral mark. You do with hydraulic system "A" on, "B" off, and vice versa, this way and that. This is many, many, steps. Then for the autopilot authority, you use the computer to perform a long test in a similar fashion, only now you might have to adjust the elevator system position sensors that the autopilot uses to control the airplane.

The aileron autopilot authority check is similar and very often the elevator and aileron autopilot functionals are combined in the same time period. The main difference is that the elevator functionals require simulated air speed and the aileron functionals do not. But the aileron functionals are tied in with the flight spoiler sensors and that gives the ailerons a little extra dynamic. These functionals require the best that flight control shop has to offer even when they work like they are supposed too. But when you have problems come up here, it really gets interesting. In flight control shop we are fully capable of performing the functionals, and troubleshooting some problems, but when it gets deep we call for Avionics to help bail us out. However, definitely in the era of the future, avionics and mechanics must become one. Many times I found a problem to be at the intersection of a mechanical and electrical system. And purely dealing with just one or the other wouldn't get it. Understanding both at the same time is necessary for efficient troubleshooting.

Of course the engines are run, after having been removed and reinstalled, and the APU is run and adjusted. Got to be careful when you run the engines, the manual says if you run both to takeoff power at the same time you can drag the tires across the ramp, even with the brakes set. So the power assurance runs are done with chocks under the wheels, and a careful operator in the flight deck, 'cause those babies wana run.

The aircraft is pressurized to see if it has any air leaks anywhere on the pressurized portions of the tube. The APU is used to run the air conditioning packs in order to pressurize, or if the APU is not running for some reason, or if the aircraft is still in the hangar, the jet will be connected to a "Huffer." The Huffer is a very large air compressor that is towed around on its own wheels. It uses a very loud diesel engine to run a very loud air compressor to pressurize the jet. Some might conclude that its real purpose is to be loud. The APU is much better. When the jet is being pressurized there will be swarms of inspectors circling all about looking for little air leaks at all the seams, doors, and anywhere there has been a hole poked in the jet.

When the jet is declared "green," meaning it has no open issues and is ready for delivery pending a good test flight, the airline is called upon to provide an aircrew for that purpose. On the test flight we will be keeping our eyes out for anything, but in particular we are checking out flight control trim characteristics, and door air leaks. In addition, there are an array of functional checks performed by the flight crew in the "real" environment of 30,000 feet. Manual reversion, alternate flaps, stall warning, floating spoilers, etc. Manual reversion means turning the hydraulics off and seeing which way the plane goes. What are specifically being tested here are the elevator tab rig, aileron tab rig, and the flap wing chord plane setting. Floating spoilers are flight spoilers that have weak actuators. When hydraulic power is turned off to them, they raise up under the force of the air pressure that is built up under the wing, steering the plane without being commanded. No passengers on this flight, save for a couple mechanics. A mechanic who works directly with the pilots for performing the tests sits in the jump seat, in the middle between the two pilots. This mechanic is usually supplied by the airline customer, and "buys" the test flight. We normally head out over the ocean, out past

Neah Bay. I love the view of the north side of the Olympics on a clear day as we are heading out.

The following is a story from a test flight after maintenance. It happens to be the last test fight the author went on while employed at Goodrich. It is typical; one of many, many, routine test flights, most of which are fairly uneventful.

Thursday, July 26, 2007

Goodrich Hangar 1, Paine Field.

It was my Friday, and the day was two-thirds
over when NXXXXX went for it's
first test flight. Mark went along for the ride.
The jet came back an hour later,
after the normal test flight time, and the result
was that it pitched several turns of
stab wheel, and it rolled six units left wing
heavy. Mark and Jerry got to work
quickly on the elevator tab issue.

As for the roll, I suspected the flaps might be
a little out of whack, as we had done
some work to the left outboard aft flap. The
inboard end of the aft flap was above
the mid flap trailing edge wedge, and
experience had shown that that is not a good
place for it to be. But first I wanted to
eliminate the ailerons as a source of the
trouble. And as I was about to check them, I
was told by my fellow mechanic that
the roll problem was only after the hydraulics
were turned off. Well, if it only

rolled after the hydraulics were turned off, it pretty much has to be aileron tabs.

Sure enough, I dropped the center aileron rig pin and checked the left aileron tab with my handy tab straight edge. It was .090", just on the top side of the range, but good. Then I checked the right tab. It was .130" in the center, too high. When hydraulics are turned off, the high tab pulls the aileron down, raising the right wing, dropping the left wing. Well, there the problem seemed to be. The tab was only .040" out of adjustment, but they are pretty sensitive. So I adjusted it down to .070", right in the middle of the range.

They wanted to try a second test flight right away, so I was asked if I would go and I said yes. We towed the plane into launch position, and then took off again. An empty '37 climbs fast, puts you right in your seat---hard. In ten to fifteen minuets we were up to altitude and we tried turning off the hydraulics again. As the airline rep flipped the switches, I felt the airplane yaw a bit. The standby rudder PCU was apparently misadjusted a little. Well, the aircraft did not pitch at all. Mark really knows how to make those elevator tab corrections. It still rolled at first, but a rudder trim correction of one unit made the jet fly straight and level. The aileron tab adjustment I had done did the trick as

> well. The standby rudder PCU had kicked it a little, but the tab being a little off the first time had apparently exacerbated the problem.
>
> All was good now, jet accepted. But now, I was hijacked. The rep wanted to take the jet to Sea Tac to pick up a new jet to be worked on, and then fly that one back to Goodrich. It was going to be a long night for me.
>
> We arrived at what I thought was Sea-Tac, but we had actually come back to Paine Field instead. The plane at Sea-Tac was not ready to be picked up, thank goodness. I got off work at my regular quitting time of 0230. Day's done, week's done, time to head home for the weekend.

When the test flight is good, the customer often takes the jet right then, but sometimes it will sit for a day or two. There are paperwork issues that take place, FAA forms and so forth, but that is pretty much the end of the process.

Besides the heavy maintenance, the repair station also serves as a body shop, and an emergency repair shop for larger incidents.

One time we had a 737 come in looking like it had been in some air-to-air combat. And in fact it had. It had been hit by a flock of large Pelicans. These are big birds with some solid parts on them, and it showed. One had slammed a huge dent into the base of the vertical stabilizer leading edge. A couple of others had almost obliterated the radome at the nose of the plane. At least one other had smacked into the inboard side of the right engine, ricocheted into the tube, and actually broke out a couple of the windows; *the carcass landing in a passenger's lap*! The author actually removed a couple of the bird's parts that were still in the cabin when the jet arrived at our station. That would really ruin your flying day! But, the jet landed safe and sound nonetheless.

One time the author got to lift a small plane off the Paine Field main runway that had bellied in. We used one of our large cranes. The fire department was all over of course, and they were really happy with us for being quick to respond and saving the day, by clearing the active runway. It was kind of exciting. Management took the credit for the action of course.

In 2007 the author got to work on one of the Air Force Two 757's. I helped to do some wing flap indication adjustments that were the result of routine maintenance. We had to sign a form saying we wouldn't look at any of the secret communications gear on board. It was a pretty cool experience I thought, and I was glad to say I did it. I could tell you what the inside of the plane looked like, but then I guess I would have to kill you…I know…very funny. Whatever?

***

By 2000, the 737 program at Goodrich had achieved a mature state, little supervision was required. In fact, less supervision usually made things go the smoothest and fastest. The workforce was physically mature as well, most of the mechanics were in their thirties or forties, and very experienced. Another aspect of Goodrich at this time is that is was no longer run like a small business. It had gotten very corporate in nature. Individual achievement was less rewarded. Now we were one big "team." The people were happier but the pace of work had slowed for some. There was less personal accountability, and the character of the worker was the main determiner of progress.

After the change of the millennium, and definitely after the airline drawdown following 9/11, the attitude at Goodrich began to decline. The reason for this was a change in management style brought in by a proprietor of Lean Manufacturing techniques.

Goodrich had thrived under the benevolent leadership of John Martin, but the Goodrich senior management wanted a business run using Lean concepts, the set of practices usually described as being pioneered by Toyota. At Goodrich, John Martin was replaced by an executive who had made a name for himself by using Lean at another division. That other division was a manufacturing arm. This executive brought with him a team comprised of people from Delta Airlines. Together they did their best to ruin what had been gained in the late nineties under Martin.

Lean is supposed to work in a manufacturing facility, where the process is the same over and over again. The process does all the work, and all the people have to do is make sure to follow the process very carefully and all will be well. You start with some raw material, shape it into a widget, assemble the widgets, and voila, a product.

The problem was that Goodrich was a *repair* facility, kind of like a hospital. You start with a fully assembled product, which is ill or needs some modification. This involves a lot of individual decisions and custom work. It is not a process that is 100% repeatable time and again. The skill and individual decision making ability of the workers are much more important than someone following a set process. This point was made again and again, but the powers that were turned a deaf ear. The new leaders of Goodrich were going to jam that square peg into the round hole at all costs, even if it killed everybody. A general trend back to dictatorial leadership and negativity ensued.

Another aspect of the Lean world, is that it is kind of "pretend," it has pretend businesses, and pretend customers. You really have to buy into it and play along for it to work. As Americans, we are always looking for the bottom line; the straight skinny. And we don't play follow the leader as well as some other groups.

At Goodrich, there was a layoff for the first time in more than a decade, and only the second one in company history. The first one was a short layoff during Desert Storm, which had shut the place down for a month.

This was quite a shock. Some disgruntled people took a voluntary severance. Some were laid off against their will. As there were no jobs to be had outside the company for a little while, the pace of work quickened noticeably by the remaining, and concerned, mechanics.

Goodrich ATS had peaked out and had seen the best times it would ever see. Some of the management that had been with the company since its early days started to get jobs elsewhere. That was definitely cause for concern, as these were Tramco hardliners, the type that didn't give up easily.

Eventually the Lean stuff kind of petered out, but the dictatorial and negative management attitude stayed. The subjective evaluation returned. With no money to give out in raises, supervision used the evaluations mainly as a tool to gripe at people. A lot of the middle-management people that had moved up under the early hard driving period of Tramco, before Martin, returned to the bad ways of the past to some extent. According to managers, problems were always the "mechanics" fault, never management. Real leadership was sorely lacking. Since the program had achieved a maturity and self-sufficiency, the middle-management types could get bored sitting around all the time without enough decisions to make. The arena of the hangar floor once again became a target rich environment for some small-minded middle-management types to massage their pathetic little egos at the expense of the mechanic. The only things to keep them in check were fear of lawsuits by upper management and the need for production.

In an apparent attempt to stem the downturn, parent Goodrich corporation management brought in a carefully selected executive to try to turn things around. This new division president showed promise. A suave French national with a thick accent, we called him "Frenchie." He was generally well received by many of the rank and file, which is interesting when you consider that he was opposite a bunch of tobacco chewing Midwesterners. Even though he was to beguile us with high-level executive rhetoric, and with presentations full of statistical charts and the like, he had started out his career as a mechanic himself in Europe. He came off as very personable to the mechanic. But after only a short time he was recalled to another division. Goodrich had apparently given up.

One concept that did remain was the idea of "continuous improvement." Unfortunately, that seemed to amount to a reason to never be satisfied with what Goodrich ATS had. And that is why most of the valuable and experienced people at Goodrich ended up leaving for other opportunities, mostly to Boeing. The charge of "rats abandoning a sinking ship" was often heard. They were just tired of never being able to make their employer happy.

Little did we know it, but the parent Goodrich corporation had sold our division to a foreign firm[5]. Even though the employees of a corporation make it what it is, especially a service provider company, they are left in the dark often times as to greater events that impact their lives.

Goodrich had an amazing resource in the personnel it had, and lost it all. A union supporter would say, "I told you so." But is that the end of it? Is the socialist union the only answer? Are the only choices a tyrannical non-union company or a traditional union company?

---

[5] See final Goodrich letter, Appendix D, p. 196.

Chapter IV

The Boeing Experience

As a preface to this chapter it should be noted that, historically, Boeing makes the best airliners. They have a reputation of having some of the best engineering resources in the world. From the airline customer point of view, they have the best product support of the airliner manufacturers. There are not too many heavy industrial products that the United States is still #1 in, but Boeing airliners are one of them.

In regards to the jet age of the twentieth century, some mechanics like the old McDonnell Douglas, some like Airbus, some even liked the old Lockheed. But if you took a poll of the best strengths/weaknesses ratio, Boeing would win. Boeing does not have a problem with products, science or engineering, but it does have a people problem. And that costs Boeing efficiency, and costs a lot of people frustration.

Boeing is of course a manufacturer, not an aircraft maintenance facility. But, believe it or not, it is a maintenance facility for Air Force One and a few other planes from time to time (customers are told to bring their check book).

However, the building of airplanes and the maintenance of airplanes are done by the same types of people, often by the same methods.

Boeing is a traditional union company. The people who physically build the plane are represented by the International Association of Machinists and Aerospace Workers (IAM). The engineering staff, the people who design the planes, are represented by the Society of Professional Engineering Employees in Aerospace (SPEEA).

Boeing was a mid-twentieth century success story. World War II built the company, producing B-17 and B-29 bombers for the military. The 1950's saw the trend continue. Boeing produced piston engine civilian airliners, but was a minor player in the civilian market. In the 1950s Boeing introduced the jet propelled B-47 bomber for the military, then the venerable B-52 bomber, still with us today. Just following was the first American jet airliner, the 707, which was to set Boeing on the path to commercial airline dominance.

The author grew up in the Pacific Northwest, where Boeing's main facilities are located. Assembly workers at Boeing were often very frustrated people in the seventies, eighties, and nineties. The old timers used to say that in the fifties and sixties, the era of the prop planes, the 707 and 727, Boeing was a regular sweat shop. The World War II generation handled that fairly well. The next generation did not. The baby boomer generation and later generations were very frustrated with their Boeing. There was a longstanding reputation of the battle and mutual dislike of labor vs. management. And there was a constant cycle of mass lay-offs and hiring sprees. The union Machinist workers had a love-hate relationship with Boeing. They would constantly gripe about the management the entire time they worked there, get laid off for a couple of years, maybe go to school for a degree that they will likely never use at the state's expense while collecting unemployment, and then pine for the day they could return to the "Lazy-B Ranch" to bang some more rivets. Even though the Machinist workers would often be frustrated with their Boeing, it was no secret that they didn't have to work too hard while they were there, and of course the pay and benefits were judged as top of the industry, or near the top. Many Machinists had no more than a high school education when first hiring on at Boeing. And many were not "professional" aviation workers per se, they were just people looking for a job, and Boeing happened to be hiring at the time.

People in engineering on the other hand, often thought of Boeing as the greatest place to be. The engineering world had much more of a sense of ownership, and therefore belonging. And, conditions for them were better. They were the "professional" class. But by the end of the twentieth century, even that relationship began to slip a little.

In the 1990s, Boeing was feeling the effects of its lost market share to European archrival Airbus. They felt the need to revamp their corporate culture, as their internal costs were deemed very high. After repeated strikes over the years, management bore a lot of animosity towards the unions. To make matters more complicated, Boeing merged with domestic rival McDonnell Douglas, a company that made a different style of airliner; rather opposite of a Boeing. Now it was simply the USA vs. Europe. In this mood the 787 program was conceived.

The author was not too keen to join the Boeing crew when the opportunity first presented itself, but with the mass exodus from Goodrich to join the 787 program, the choice seemed obvious, if not necessary. The veteran mechanics of Goodrich thought they would join the Boeing flightline, which was judged an easy job for an experienced, well-rounded mechanic, and "take the money." A few years like this to cap off one's career wouldn't hurt, even if one were anti-union at heart. Besides, you could reason that you would produce some good work, and not rely on the union at all for anything. After all, Boeing was hiring Goodrich workers ostensibly for their performance ability.

The first couple of weeks, or in some cases the first couple of months, the new 787 hires were sent to work at the legacy (777, 747) airplane stalls on the flightline (a stall is a work center on the flightline, sort of like a gate at an airport operation) to "shadow" or see what it is like to work out there. They jumped into the extensive required training at this time also. Management and co-worker alike treated the newcomers well. Everyone thought, "this is great."

The flightline at the Boeing factory is different than its airline equivalent, it is the last stage of a manufacturing process. But, it does involve dealing with problems, fueling, running, and launching aircraft. The "traveled" work is dealt with on the line. Traveled work is that work, which in the factory was not accomplished when it was supposed to be, hence, traveled.

Historically, the most experienced functional test workers from the factory fill the flightline jobs. But, for the 787 a change in philosophy was made and Boeing wanted licensed and experienced airline mechanics. They hired quite an impressive team from around the airline world. Mechanics from American, Delta, Northwest, United, several smaller airlines, and just about every highly experienced Goodrich mechanic, lead, and QC they could get. The two largest groups were Goodrich first, and Northwest Airlines second. These two groups formed the majority of the 787 flightline group.

A group of five former Goodrich workers were sent into the factory in the summer of 2007, of which the author was one. Many more would find their way into the "Big House" in the next several months.

My first impressions of the factory were mixed. I had heard what a mad operation the 787 program was, that it was really involved. But when you went into the hangar bay, all you saw were groups of young people sitting at large tables staring at laptop computers. No one was working on the plane. Strangely, this did not seem to alarm anyone; at this early date.

A group of guys from Goodrich that had hired into flight test had already been with the program for almost a year. We all knew each other and it was a comfort to have such a familiarity. We all thought the same thing "wow, what a different sort of operation."

Right away my attitude was like "well, lets do it, lets get to it." "Lets get this done so I can get back on the flightline." But I quickly learned that it was not that simple, and this was going to be a different world. In the next year I would suffer frustration like I had never suffered it before.

The Boeing senior management was not happy with the level of factory efficiency in the last decades of the twentieth century. There had been talk of Boeing leaving the Northwest for one reason or another over the last decade.

Like everyone else these days Boeing is a purveyor of Lean. They like to tell themselves how Lean is improving this, and improving that. But, like most other American companies, it is not as easy to see these results when on the inside. Those old prejudices, fiefdoms, and inefficiencies die hard. And most companies tell their workers to have one standard, and reserve another for themselves.

Boeing had decided that they were going to throw everything out they had learned in the last century and start over. They would employ all new (and mostly young) people, un-jaded by sloth and old values. Almost no experienced, old-guard, union people were allowed to play at first.

The plane itself seemed marvelous when viewed in computer simulations, and in the training classes of its systems. The structure of the plane was impressive. The plane was made from new materials—composite airframe with titanium hardware. Almost none of the traditional aluminum. The concept was ingenious, no aluminum to corrode, and good gas mileage. Done deal for the airlines. No wonder they had so many orders. Strangely, even though they had so many orders, Boeing still felt the need to "market" the pane all the time, to count their chickens before they hatched.

There was new technical data, all laptop computer based; the paperless ideal. New terminology, for absolutely everything. Even the common hardware that airline veterans were used to had different part numbers now. And of course the big one: The plane was not really manufactured in Everett, but all over the world, and the higher assemblies were merely to be joined together in Everett.

New processes, for everything. Boeing thought they could engineer a process to build the plane that they could literally stick *anyone* into; and out would pop a plane.

The process alone, the Lean process, would build the plane. They made all these changes at the same time. In with the new and out with the old. Throw out the baby with the bathwater.

They hired all new, mostly young, people for supervisory positions and workers alike. They wanted a culture change. They wanted the end of the Legacy Boeing culture. Old was bad.

For many of us coming from Goodrich, we sided with the company. We thought the IAM culture of Boeing was rotten as well. We thought ourselves morally superior.

One thing that struck one about the 787 operation was the mobility of everything. Everything was on wheels, in sharp contrast to the legacy programs. Portable manufacturing. This operation was clearly designed to be capable of being conducted anywhere. After all, the major assemblies were flown in from the four corners of the earth on the huge, specially constructed, 747-based Dreamlifters. They could fly those parts just as easily somewhere else.

Working on the nose end (that is the 41/43 section, for you Boeing aircraft types) of plane number one, we were literally at the cutting edge of this historic project, I guess you could say. Most of the time we were just not able to do anything due to lack of parts, or technical data that was fraught with errors, confusing terminology and document formats (even for veterans, nay--especially for veterans), inability to get answers from engineering, or lacking some tool or something. If you did anything it was usually making a vain attempt at fixing a problem of some kind, hardly ever would you be actually able to build the jet a little. No one was very helpful at getting a job done around here. The place seemed to be geared towards finding reasons to *not* getting anything done. Truly amazing with all the hype and spotlight. The processes had absolutely everyone stumped. People were always quick to point the blame down the hall. "It's those people over there, they are the problem."

A lot of the workers at Boeing act as if their individual efforts do not make much of a difference in the overall outcomes. As if they are just a number. They behave as if they are actors in a play.

A couple of managers tried to make progress, but were eventually swallowed up. Most of them just ran their yaps and pretended to be on top of things. "Plenty of overtime available to come in and do nothing some more." How anyone could believe a word they said was hard to believe in itself. The managers could sometimes say some very intelligent things, but after a while it seemed as though, they too, were merely actors in a play. "Who was the puppet master?" I wondered. The veterans coming in from the outside could hardly believe their eyes. But, they were caught up in it as well, and had become part of the conundrum.

One thing that jumped out at the aircraft maintenance veterans, was that this was actually more like aircraft *maintenance*, and less like aircraft manufacture. It was a lot of dealing with special isolated problems. And not so much a back-to-front, step-by-step assembly process.

As for all the new people they hired, the results of their unguided inexperience were to be seen everywhere. It was a completely unrealistic expectation that you could take a bunch of people who have never worked aviation before, send them to classes for a couple of months, and expect them to build a twenty-first century jet un-aided. *Re*work was to be as common on this project as work. Many of them were actually quite good, however, and quick learners. To give them credit they did very well under the pathetic circumstances. But it was very much a case of the blind leading the blind. And the results were telling.

That is not to say that the problems started in the Everett factory though. The problems started with the different vendors, and the work, or lack of work they performed. The planes are built in sections around the world. This company makes a 41 section, that company makes a 48 section, this other company makes wings, etc. Boeing's problem was that they only had one vendor make this assembly, and one vendor make that assembly. They gave ownership to the different vendors, so they could not have a second vendor making a particular assembly and therefore have some competition. Consequently, a planc was like a convoy that had to travel the speed of the slowest ship. This led to another metaphor: It ended up a train wreck on the factory floor.

The Boeing legacy programs were designed to control the build of the plane from one end of the factory to the other. Control, control, try to keep things under control. Keep the amount of failures small. Unfortunately, the amount of failures on the legacy programs were never small, and that is why the new upcoming 787 program was changed so extensively in the first place.

But this was a whole new level of bad. The 787 program had small failures on a massive scale, and overwhelmed the usual Boeing system to cope with failure. The Boeing process to cope with failures was slow and tedious. Adding insult to injury was the new inexperienced people, the unhelpful attitude of the supporting staff, and the inefficiency of the supporting vendors. The result was gridlock.

Boeing is supposed to be committed to Lean. One of the tenets of Lean is to have the customer-provider relationship represented in the workplace. In an aircraft maintenance or manufacturing environment, the worker with a job is a customer. That person needs to complete that job to satisfy their customers: Boss Man (who represents the external customer), and the quality organization. It is generally accepted that micromanaging is not an effective management method given a detailed job scope. The worker needs to be free to be able to complete the job using the resources that are available. But, the worker needs to be accountable somehow.

The worker is then a customer. The worker's providers are parts procurement agencies, engineering agencies and their liaisons. Also special tooling services, and maybe special process services like structure bond (special composites repair), non destructive testing, and special machining services.

In your everyday life, you may need the services of an accountant or lawyer. These are highly trained specialists that deal in information. If you solicit their services, you are the customer, and they are the provider. They work for you. They are providing a service to you. They may be more educated than you, make more money than you, but for this transaction, you are the customer and they are working for you. This is accepted.

Boeing manufacturing culture doesn't get this concept. They like to operate on a subtle class system. They like to keep the plebs in the dark. In this aspect of Lean operations, Boeing gets an (F). The runaround one receives while trying to accomplish something is off the scale. At the very same time, there was no worker accountability at all. Result: (no worker accountability) + (no parts availability/no parts procurement efficiency) +

(poor liaison engineering customer service)/(poorly written job dispositions/job cards) + (weak, visual oriented management) = zero, nada. A recipe for disaster.

Being true to form, the author started to send emails and letters to management officials pointing out the perceived problems. Not much response. Hands thrown in the air.

In a computer simulation, the airplane was supposed to be built in mere days, snapped together as it were. Two years later, we would still be trying to "snap" one together.

The whole thing seemed, after a while, like a program designed to fail. You could not have engineered a better program for failure if you tried. It was a Tower of Babel.

I was working with an MT (Manufacturing Technician, the largest group assembly of workers) for a period of time. She was a Boeing veteran of the past and was very helpful lending me a hand to understand the Boeing process. One day I was lamenting how fouled up things were, and she replied, "It's always been this way." Wow, that's a depressing thought, I mused.

The events of one particular day came to characterize my perception of Boeing, from that time forward. I had been accustomed to not being able to accomplish jobs for one reason or another: I couldn't understand the data, no parts, no correct tooling, etc.

I got a job installing a part that had a filler block sandwiched between two parts. The purpose of the filler block was to take up space, no big deal. Someone had previously worked the job, and drilled the filler flipped 180°, causing the holes to be in the wrong place. We needed a new filler. It was a flat piece of 2024-T3, .140". Common as dirt materiel in the aviation industry. It is like saying Doug Fir 2x4 stud in residential construction.

At Goodrich, I would have gone back to the warehouse, "bought" some certified materiel on my job, fabricated it, treated it with alodine, primed it, maybe have QC "buy" the fabrication, and then installed it. Probably would have taken an hour at the most. No big deal.

I said to my MT coworker "we can do this." I knew it would be more challenging here, but in very rare Boeing 787 fashion, my job gave me references to make the new filler: materiel specification, and instructions to alodine and prime. Alodine and primer were to be found in the immediate area. So all I needed was materiel. Where do you get materiel? That took awhile. I was sent to the coordinators crib, a local "store" for "stuff." They carry pens and booklets, and some aircraft hardware. They did not have the correct certified materiel. I asked this person and that person. Parts people, management people. Finally, after much ado, I found out that there was a place to get materiel in the extreme other end of the factory, about a half mile away. So, I trundled off to never-never land to get some metal. Well, after finding the place, which was not too easy, I talked to a guy that looked half hung over and asked him if I could get some materiel. For a moment it looked like we were finally in business. But no, when all was said and done he could not issue metal to a 787 job. So, back I went. I talked to the parts ordering person again, and she told me that Boeing was trying to get out of the business of making little parts, and that the little flat piece of metal would have to be alodined at this facility over here, then primed at that facility over there, and then we would get it. But she didn't want to order the metal anyway. She wanted me to wait for the part, which was on order, and was going to arrive heaven knows when. I had tried my best to push a simple thing through, and failed. Another wasted day, one of many.

After all this...I reflected that here I was in the largest building in the world...which was an aircraft factory...and I couldn't make a simple, flat, aluminum part! This experience was endemic of the greater 787 environment. At that point I pretty much gave up.

Shortly later that first group of five flightline newbies, who had been working separately for different factory managers, was absorbed into one large 787 group from the flightline. They were all working in the factory for their flightline managers, but still under the auspices of the factory management. Our flightline management treated us very well, but put almost no expectations on us. In this mode the group was largely hogtied from making a real impact on the work schedule. The presence of the flightline group seemed to just piss-off the factory management, and they spent a lot of time whining about the band of "yellow jackets," so named for their yellow ANSI approved reflective coats.

During this time period the author got to know several Boeing veteran "legacy" workers. To my surprise, many of them were more skilled than at first expected, and with varied experience. Many of them were more capable people than I had expected coming in the door. Eventually I had to admit that many of the Boeing legacy veterans were not so bad of workers after all. I was conversing with one of them one day, lamenting the tedious operation around me. He, paused, reflected, and stated: "this is not the best this company has to offer," "this is not the total resources of this company," and "I'm not sure what's going on here?" That made me wonder. Later on, burned out with the 787, he found another job with Boeing at another facility.

Another thing I noticed; some of the factory managers were a little screwy. They could be abrasive for no good reason. Even a little psycho. They would feign being very polite, while at the same time coming across with an abrasive vibe. Sometimes a little nasty comment from out of the blue.

In fact, a lot of the management at Boeing reminded me of the Keystone Cops, not everyone, but too many. Running around bumping into each other, making a lot of noise, doing a lot of complaining, and getting absolutely nothing accomplished. And in the final analysis, they were most concerned with how things looked, perceptions, not how they actually were. And often times their idea of what looked good was even bizarre.

This was a very politically correct environment, no cursing someone out, unless they were your buddy. If you dissed someone, you had to do it with slight of hand. This created a strange environment dealing with leadership, a formal uneasiness. No one really being open or honest with one another, everyone pretending to like and respect each other, whether true or not. It was like an invisible nanny or Sister was present to whack you on the head if you stepped out of line (I think she's actually called H.R.). Another barrier.

During all this time, on the 787 program we had made no jets. However, on the 777 program right next door, they had been spitting them out every week. You could actually see the assembly line moving. Maybe it wasn't perfect, but it was working. It was getting a little harder to be skeptical of the Boeing legacy crew. And it was becoming a little easier to see flaws in Boeing management culture.

But then, the strike build-up started to gear up. This was going to be quite a ride.

This was the first time that many of us had ever worked for a union company, or had to deal directly with a strike. The propaganda surge reminded of me of the efforts of Joseph Goebbles, propaganda minister of the Third Reich. Slogans, buzzwords, slanders, etc. There were to be pep rallies, marches, and so on. Months and months worth of them. Great and long marches through the factory. Some veterans said that they were the longest marches they ever saw. And the issues of the strike? I think it was all about not enough bubble gum machines in the factory or something, I can't even remember. Lots of people walking around in light blue shirts, "Contract 08, No Time to Wait." Get on the bandwagon, it's the thing to do. Oh, boy, puts a lump in your throat. Gives you a warm and fuzzy.

To most of the new MT's it was just a game, like High School sports. It had a carnival atmosphere.

Then came time for the strike vote. Strangely enough, when it came right down to it, people just wanted a month or two off. They just wanted the time off. The contract looked as though it was going to get resolved at the last minute after the strike vote, and many people were just livid. They were pissed. They had plans, they wanted the time off. The union needed the strike, so it happened.

The whole factory shuts down so school could let out. Well, you've got to strike every now and then to show them who's boss, right? Whatever.

Two useless months roll by and it was over. Useless strike. Nothing really proved. The us-and-them game was on full swing from that point on though. When we came back to work, the management attitude was definitely no-more-mister-nice-guy. Union negotiated ten-minute break, yada, yada.

Several of the flightline group didn't really support the strike, but we went on it anyway. When we came back, we weren't really all that surprised by the new company attitude. Even sympathized a little.

But, the upper management started to lose us a bit at this point. They blamed 787 delays on the strike. The strike had no effect on 787 production in reality. They were so behind on parts and engineering issues it should have been a *benefit* for them to have us all out of the way. But no, they sat on their hands during the strike, and tidied up the factory. The lack of 777 production on the other hand definitely put the company behind.

After a few months with the flightline group, it became obvious that Boeing had indeed hired quite a special group of people. People of all races, and from around the world, we were united in thought and interests by our shared profession. For the most part, we liked each other. It was a fun club to be with. We were in a very frustrating environment, but we got along with each other.

We made it through the holiday season, and then our flightline management convinced senior management to let the flightline group have jet one all to ourselves, away from the main factory operation. Our own quality people, our own mechanics, our own managers. And we even got a captive audience support staff that couldn't just blow us off easily.

It was still a challenge by our pre-Boeing standards, but this did the trick. For the first time since we had been there--things started to happen. Experience did in fact make a difference. The experienced Goodrich, Boeing legacy, and airline mechanics knew how to "get things done," even in spite of the restrictive processes, to push them through. If a mechanic takes ownership of a job, he or she uses their skills to network with quality, parts obtainers, and engineering liaison services to sort out complicated problems. You only use your management if you really have to, because that tends to put people on notice. Cooperation works best whenever possible. But people have to be self-motivated for this to work. Building a new jet, the work has to be done to the highest standard. That is a challenge in itself, when someone less than well skilled has worked a job before you. You have to sort out the mess. That takes skill, tact. A manager cannot micromanage such transactions, they will only serve to slow things down in most cases. Some managers, however, want to make the noise and get noticed, because they of course want the credit for "making things happen." However, the successful manager is the one that can somehow motivate *others* to get things accomplished. That takes a lot of leadership skill. These type of leaders are always in short supply, unfortunately.

The first week in our flightline-only environment you could see a difference. In two months we had made more than a year's worth of progress at the previous factory rate. But during this time, as we did better and better, management was still harassing us with "visual" oriented B.S. They just could not get out of this visual, perception-oriented management. Just too much of a spotlight I guess. But this external visual approach to management caused untold hardship, consternation, and lack of real progress throughout the entire program.

Despite all the pitfalls and roadblocks of the past two years, after another couple of months had gone by, it became obvious that the blue and white jet was finally almost built! Never in our lives had any of us spent so much time on one single airplane. But at least now the end was in sight.

Then came the fateful day when the plane actually rolled out to the flightline. It seemed like it would never happen, but it did. And it was mostly due to a bunch of AMT's. Veteran Guys and Gals from Goodrich, the airlines, and Boeing legacy programs. And it should be noted that the few Boeing legacy mechanics we had, being more familiar with Boeing processes and procedures, did especially well throughout this time period.

Zoom seven years ahead and the 787 manufacturing process has improved, but not as well as desired and the 787 still has failed to generate an overall profit. The analysts make dire predictions that the startup costs were so great; the program may never see a profit. If it were the only model Boeing had going, it would be ruining the company, and Boeing would be another American corporate statistic. This calls to mind the earlier example of Outboard Marine Corporation, whose labor problems led to outsourcing, which in turn led to OMC's final demise. The basic concept of the 787 manufacturing process, to have subassemblies built in remote locations and then brought together for final assembly is a proven concept—it has worked in the past. That in itself was not the issue. But initially the partners failed to deliver. They had made commitments they were not able to fill. This was due either to their inexperience or culture—their corporate culture or the national culture where that firm was located.

Perhaps if Boeing had a group of vendors to compete on producing individual items instead of relying on a single source, where that source has a little too much leverage over the big pie? Or is there another way? That is the subject of the next chapter.

One thing is for certain, there will be more international competition for Boeing in the 21st century and doing business in a late 20th century fashion does not seem a viable course—and that goes for the union as well.

It is the opinion of the author that Boeing has the people it needs to succeed, they just need the right attitude. The culture tends to be a selfish, me-oriented one. Perhaps it is a function of greater modern America.

The Boeing worker tends to keep useful information to their self. They want to have one-up on the other person. Many are very intelligent and capable, but tend to be cynical and not interested in the overall progress of the work and the company, merely their own personal success. They do not necessarily see themselves as a part of a larger community with a responsibility to that community. If there is the typical layoff, they want to be the one that is retained because they are special and the operation can't go on without them. There is a sort of elitist culture, those in the know and those not in the know, which is very counterproductive to the overall mission.

However, the company, like any other, stresses protecting its proprietary information to maintain its competitive advantage in the marketplace. The individual might then consider that they are protecting their own "proprietary" information, their expertise, to remain competitive in their own labor "marketplace." This does not help the company's goal of producing a product however, and is an impediment. Ideally, within the company information would be freely distributed in a spirit of selfless teamwork towards the end of producing the product. Human nature would seem to tend contrary to this spirit of teamwork, however, and the selfish nature prevails—self interest.

There seems to be a double standard at play, which tends to hinder overall progress. The company wants to protect its proprietary information at all costs, while at the same time expecting its employees to be dutiful, selfless, and all giving. Of course many employees are not selfless and all giving so this creates inefficiency, which slows the entire organization.

On the other hand many people feel as though they are a meaningless digit, and are talked down to and slighted by management. There is the perception that operations are micromanaged from a high level, which creates a perception of chaos from the view of the lower levels and a sense that no one must be in charge, or no one with a clue. This of course tends to promote worker apathy and a sense that there is nothing an individual can do to improve things. One can only play along and take the benefits offered, reinforcing the "me" culture.

In the final analysis, one must surmise that this condition must be a function of the structure of the organization and the management/union relationship. To change this condition for the better therefore it would be necessary to change that structure and management/union relationship…or perhaps move beyond the traditional management/union relationship altogether….

Chapter V

The way we Ought to Work

What motivates people to work? The stick, or the carrot? What about neither? In the one psychology class that I ever took, I learned that there was only *one* fact in the "science" of psychology. That would be the so-called Pavlov's law, which states that a positive stimulus works better on an organism to illicit a response than a negative stimulus does. Simply put, if you want the dog to do something, it works better to feed it than it does to kick it. But we see in life that both do work, kicking the dog works, but feeding the dog works better. This is true, this is in fact how they actually train dogs. What does doing neither do? Not much really...as they say, "neither here nor there."

Leadership styles

How does this apply to our world? It would seem that magnanimous leadership would be more effective than harsh oppressive leadership. Again, both work, but I would have to say that the magnanimous leader/leadership gets the best results. The magnanimous leader tries not to anger those below them. The oppressive leader tries not to anger those above them. In my experience, it is easier to be an oppressive leader than a magnanimous one. The magnanimous leader gets the best results, but puts in the most effort. The middle of the road leader tries to keep a low profile, and tries to not anger either those below or above. They get very little results.

Middle of the road leaders are the most common I would say. Mediocre results are the norm. Oppressive leaders are the next most common group, the guy who kicks people in the ass and gets things done. Unfortunately this kind makes a lot of enemies along the way and usually faces some sort of payback at some point. Think of the peasants storming the castle gates with pitchforks. Middle of the road leaders sometimes get this reaction also because of the chaos they create by inaction. Think of the Russians killing the Czar. Then there are the smallest category of all, the truly magnanimous leaders, the leaders who inspire men to do their best freely. I'm sure you can think of a few examples. Such a leader can do much great work in a short period of time. The trouble is, they tend to be few and far between, historically. Maybe one out of five. So that gives us one step forward, four steps back. This is why monarchy does not work well. This is why American corporate structure does not work well. This is why our industries have been faltering for decades. Our management is a monarchy, and our labor structure is based on a communist model. I.e. hourly labor controlled by large unions. Equal pay for everyone no matter what kind of work they do. This arrangement has toppled several of our American industries that used to enjoy a monopoly on the world scene.

Labor structures

If we travel far back into the history of western civilization, to the late Roman Republic, we see that groups of skilled tradesmen formed associations known as *collegia*, for their mutual benefit and protection. These were associations of small businesses that made use of mostly slave labor. Rome fell, and in the Dark Ages these associations fell away, as did almost all of Roman infrastructure, for a time. But then in the medieval period we see associations of tradesmen form again and become a dominant economic force in Europe. These associations were known as guilds.

The medieval guilds were associations of merchants and trades people of a particular craft referred to as artisans. These were times when an individual had very little power. A group of smiths or bakers in a small locale, such as a village, would band together in an organization for their mutual benefit and protection.

These were small businesses. The owner of the business was a master, and was considered an expert in his trade. He had to have proven this to the body of the guild members, or at least to the leadership thereof, with a masterwork that he had presented to them—this was his qualification. Now he could run his own business as best he could, within the confines of the guild's rules, which were many.

The master achieved his position by first being apprenticed to a master when he was just a boy. He served in this apprenticeship for several years, working for room and board only. When the term of his apprenticeship was up, and the master felt he was ready, he could solicit to the guild council to be granted the status of journeyman. Now he could go out and work for any master and also earn wages. To become a master himself, he would work on his masterwork on his own time.

Pausing briefly, in chapter one we saw that a modern skilled mechanic starts learning his or her trade as a child. It really takes more than a one or two year program to pick it up. In the middle ages, one might have to be an apprentice in some crafts for a decade, or even more. A smith would be apprenticed for six to eight years—and this was to learn making horse shoes, sickles, swords, and the like. An apprentice received individualized training from their master, and had to prove competency to the master's peers. In this way the quality standard of the guild was maintained, and therefore its value. Our modern idea of a one or two year vocational program seems to be born out of a twentieth century world war era government inspired need for quick training programs to fill thousands of needed skilled jobs quickly. These casually trained people are then supervised by persons very often not skilled or well versed in the work being performed. This modern way is not a balanced or comprehensive approach, and not suited to an enduring quality culture.

This was of course pre-industrial Europe, and all crafts were hand made. This guild system ensured a high degree of quality, but things were not cheap as a lot of effort went into making them. The guilds did not let members compete on price either, only service. The prices for everything were set by the guilds. The standards by which things were made were set by the guilds. The members of one guild were not allowed to perform the activities of any other guild. The guilds were monopolistic and controlling. The guilds regulated everything in the name of stability and security. In fact, by the late Middle Age, the guilds had become quite stifling in their effect on industry. Apprenticeships were hard to come by unless you were the son of a guild member, and it was difficult to become a journeyman. The guilds were perceived as stifling to technological innovation and business growth, and had become unpopular to many. Governmental pressure started to impinge on the power of the guilds in some places by the time of the Renaissance and Reformation in the sixteenth century. This was the case in England.

Lets talk about the modern labor unions for a moment. The union organizers of the 1930's, when the aviation branch of the IAM was organized, were said to be communist in their political beliefs. We have all heard of the communist ideals. Equality for all, lack for none. Did it ever work that way in the Soviet Union or China? Apparently not.

Lets talk about the world's first communist experiment. This was the story of the Pilgrims who came from England to America in the early seventeenth century and founded our country. They were financed by a company in England, and that company expected a return on its investment. They were a community that was supposed to generate an income for a company, as well as for themselves.

These were some of the most sincere Christian people the world has ever known. They walked the walk and talked the talk. When they founded their town, they determined to share everything with each other. They believed that is what Christ was commanding them to do in the Bible. They were serious Christians; they were going to put their faith to use. So, they shared everything with each other the first year in their village. It did not work well. This woman did not like washing the socks of the guy down the street, that guy did not like chopping that woman's firewood. They had a lot of want and not enough plenty. Sound like a certain Soviet Union we used to know?

Well, the next year the town leaders decided to change things a little. They allotted each family their own plot to farm. They still functioned as a community, helping each other; they still had their charitable Christian values. But now they were much more industrious, as they were working for profit. This was said to have the effect of putting all hands to good use. This resulted in a year of plenty, and most years after as well.

American business operated a little freer from governmental control than its European counterparts, and grew rapidly. American products were mostly more of a practical nature. In America we wanted the widget that worked good, not necessarily the one that was decorated fancily, like the masterpiece of old.

Next comes capitalism in the eighteenth century, and then the industrial revolution. In capitalism, private investors--the entrepreneurs, set up factories and various large assets, hire wage labor, and seek their profit. Government is kept at bay as much as possible; and ideally simply makes sure that there is a level playing field for business, no monopolies, and everyone keeps their word as to business transactions. The free market will provide prosperity by virtue of its design.

The key proponent of this Capitalist system was a guy named Adam Smith of Scotland, a Christian academic and philosopher. He wrote his philosophies in a series of works called the *Wealth of Nations*. The author has no formal economics training, but some things just reach out and grab you when you read them. That is the case here. Smith contended that personal interest for profit was the primary engine in economics. That really doesn't sound so strange when you think about it, but the concept is often muddled in today's rhetoric. It is interesting to note that Smith thought America was going to be a great nation because of its ability to embrace this new Capitalism.

The investors of this system would be the wealthy middle and upper classes, who would hire the uneducated masses for wage labor. At that time, few people had access to education. The poor people needed guidance.

In the industrialized capitalist economy, soon people were working in larger shops and the small business began to decline. Mass production methods were pioneered, and the apprentices and journeymen of the small shops went to work in larger factories, the master became their foreman. Together they worked for the capitalist investors, the businessmen. The master was no longer the businessman; his boss, or his boss's boss was. But in the capitalist industrial world, the economies of the western world mushroomed.

As the capitalist company gets larger and larger, it assumes an army like structure, and just like in the army, a great chasm emerges between the top and bottom of the large organization. Many levels of managers tend to cloud communications both ways. And, the owner holds all the cards in relation to the workers. He gets the money first, and then pays them. He has a kind of monopoly on them, he has control. The company becomes likened unto a kingdom. The worker's only choice is to find another employer, another kingdom, if he or she doesn't like where they are at. And this is how worker conditions and pay developed.

As a counter to capitalist thought, European atheist intellectuals in the early nineteenth century develop the principles of communism—Karl Marx and the convention that was the First International. This becomes the basis for the union movement of the nineteenth century. It is a plan for a new world order. The International Workingmen's Association (IWA) is formed on this platform and will spawn many unions including maritime unions. The fuel for the union movement rose up out of worker conditions in the nineteenth century. Large companies such as the American railroads were known for not treating their workers with much respect. Smaller firms such as ship owners were often harsh. This cannot be denied. The laborer on land and sea did not have it easy. The owners of their companies were known as greedy pigs, getting rich off the backs of the ordinary worker, or at least working them to death in trying to make the money. A great chasm opens up between the rich capitalists and the lowly wage-workers as a result of the size of the companies. One did not understand the other. The companies grew to represent the feudal kingdoms of old with their tyrant rulers.

In the late nineteenth century communist thought grew in Russia. In the same period of time the labor union movement got into high gear. The people of Russia were poor and did not like living in squalor while people like their Czar lived in such luxury. So, during World War I they had their revolution and killed the Czar. They went from one extreme to another politically. In the same way the labor unions overthrew the power of their robber baron overlords. This union labor movement seemed to work for half a century when people were a little simpler, and had far better traditional work-ethic values. But often the people of the union found they had a new master in their union. The people of Russia soon found they had exchanged one harsh master for another even worse. Strangely, the one thing you did not have to do in Soviet Russia was work hard. You just had to know who was boss, and everything would be o.k. In a union company you do not have to work hard, you just have to remember the union is boss. In Soviet Russia the lack of efficiency caused almost everyone to be without everything. In America, the lack of efficiency in a union company causes high prices for American goods, and eventual loss of jobs to foreign countries. So in the end, no union job at all, work is all gone.

The reason that traditional union companies are inefficient is that the labor people, the people doing the real work, are paid *by the hour*. They are essentially paid to be "on the clock" a certain amount of time. They are not paid to produce a certain quantity of work. The unions very specifically make sure that this is the condition. Then, these workers are protected from "harassment" by their management when asked to produce more. So, there is not much personal accountability. The moral fiber of the union worker is supposed to be the driving force of efficiency. This assumes that people are good, and will do the right thing.

But, Christianity and the Bible teach that man (and this category includes *wo*man) has a bent towards evil, a sinful nature. He will not always do the right thing. If you make it easy for him to cheat, he will likely cheat, and especially if he does not face much in the way of consequences. We see much more evidence of this in the world than for the opposite. That is why we have police and armies. Until the end of this age.

The founders of our country, one nation under God, realized these truths and formed a government that best managed our sinful natures. A government with checks and balances, where hopefully no one gains too much power or leverage. That is why the government breaks up monopolies in industry, because it does not serve the interests of the nation to have one man, or corporation, control an entire industry that we all depend upon. But men will lust after power still, and always have.

The modern corporation came into prominence in the nineteenth century. It was a legal means to provide limited liability to a group of investors. The corporation is a legal and distinct entity. A public corporation is not owned by one person, but many.

At the top of a corporation is the democratic organization that is the body of investors, or shareholders. The investors elect a board of directors to run the business, who in turn appoint a president, or CEO. The shareholders basically turn over control of the company to these top executives that they have appointed. This executive control is said to be particularly the case these days, as shareholders of a modern American company often number in the hundreds of thousands.

Below its investors, a corporation is just like an army, with a general at the top, officers below him, and enlisted men to do the work below them. It is a chain, a chain of command. It forms a pyramid structure, general at top, workers/soldiers at bottom. It is organized into regiments, battalions, companies, and platoons. It is said that it has to be this way. But does it really?

If we compare a corporation to a form of government, it is a monarchy or dictatorship. All the people in it answer upwards, to their boss. What type of states last the longest, and grow the biggest in human history? Monarchies, kingdoms, empires? Communist dictatorships? Or democratic republics? Ancient Rome was a republic for the first half of its existence, then it was an empire after that, but it still had a senate. It is the biggest and longest lasting state the world has ever known. It is interesting to note that Rome grew five hundred years as a republic. Then it became an empire, a barely limited monarchy with Caesar, and five hundred years later it had weakened to the point that it fell. The United States is patterned after the republican period of Rome in large part.

Several prominent city states of the ancient world were democratic in government as well, such as Athens of Greece and Tyre of Phoenicia, places that contributed greatly to western civilization: culture, technology, and philosophy. The democracy is the place where forward movement occurs; it can be shown again and again. Dictatorial control leads to ruin, again and again. But dictatorial control is the norm throughout history nonetheless. Monarchies and dictatorships rise and fall again and again. People want control. And most of history is stagnant in terms of progress: scientific, cultural, and otherwise.

Corporations are economic institutions, they exist to make money. It's all about keeping that stock price up. Lets look at comparing a corporation to an army. Armies are not normally known for their economic prowess. There are some exceptions, notably the Roman army, which built many things that stand to this day. But usually an army is only a good *consumer* of resources, and not a producer. And, an army requires some kind of cause to be honed to a state of efficiency. An army without a cause tends to fall into decay.

I was in the U.S. Air Force. It is a not-for-profit organization, and I saw many examples of that when I was in. I joined in the post Viet Nam era military, and drugs and corruption were commonplace at the time. But by the time I got out, it was in the Desert Storm era, and it was a whole new Air Force. The people of the Air Force now had a cause, defending the whole world from bad guys, for doing their jobs and it seemed to have a dramatic effect all around. I think a strong leader like Ronald Reagan had a lot to do with it. But without a cause and purpose the military degrades. The military is necessary and is what it is, but it is not the best model for the civilian economy. And another point that was made to me many times when I was in the military is that it is not about making money; in the military it is all about capability and readiness to fight. They will tell you they are about military results and not about making money. They are about *spending* money. So, patterning a business after their model does not seem all that wise, does it?

We live in a democracy, a republic. Not a Soviet dictatorship, not an empire, not even a socialist state (although sometimes it seems that we are getting there). We have elections. We vote for our mayor and our President. Elected leaders are supposed to work for us (I know, just humor me), that is the big difference. I am not a *subject* of the king; the king is subject to us as a whole.

People in small business often see the world a little differently than those of us at large companies. If we are in a small business in a town, we likely belong to the chamber of commerce and are involved in our economic world around us.

But if we work at a big company, we feel like we are in an army or dictatorship. The company dominates much of our life. At the company we do not have free speech, we do not have the right carry a gun in most cases (an American 2nd Amendment Right), and we are in general less free than when we are not at work. At work we step into the world of the company, be here at this time and do that at that time. For the majority of people who work by the hour, we are paid to *be* somewhere at a certain time. We are not paid to *do* a particular service, but are paid to be somewhere at particular times of the day, and then to do what our management requires of us, if they can make us.

By contrast, if you have a small business in a town of small businesses and you need your car fixed, you take your car to the mechanic and pay him to fix the particular problem. When it is fixed, you pay him. You do not hire the mechanic and put him on your payroll, set him up with medical benefits, tell him when to show up and where, and then set expectations for getting your car fixed. You think you have problems getting your car fixed now!

One of the big problems in a corporation is the filtering of information up the chain of command, and down the chain of command. It tends to get distorted both ways. A vehicle of the corporation's inefficiency due to its military-type pyramid structure, and large size, is the subjective performance evaluation. The subjective evaluation is often seen as the most important tool in a corporate body to keep people in check. People need to be held accountable for their actions, and I would wholeheartedly agree with that. But the way it works in corporations does not achieve the end that it is supposed to. People are often not evaluated "fairly." Personalities conflict, people have differing views on what constitutes real efficiency, and what is merely "show." "Visual progress" is not always "real progress." In the reality of the rating of performance evaluations, it comes down to the networking of friends and like-minded people.

People will often say of any system, "if people will just follow the system honestly, it would work." If only this…if only that…if…if…, but people are people, and they have not changed all that much over the millennia. We just have more information available to us now. The systems that we have often rely on absolute integrity among the members of the organization for that system to work as designed. If people are biased, just flat out wrong, jealous of others, lazy, racially prejudiced, or a host of other sins, the system will not work as designed. The answer is to have the system that correctly addresses human nature, then, and only then will it work. A.k.a, the American democratic republic system.

In the Air Force of the 1980's, they had a performance evaluation system that had graduations on a scale from one to nine. Level one was poor, level nine was essentially perfect; walks on water. But due to imperfections of people and their perceptions, a rating inflation had taken place. The top of the scale had become the "normal" rating. In fact, if you compared the situation to a traffic stoplight, a nine was green, an eight was yellow, and a seven was red. If you gave a guy a seven, you were basically trying to fire him. But, by the system's design, a seven was a very high rating. In fact, most people were not even a seven if judged exactly according to the system's criteria. So, when they evaluated you, they would have to make up a bunch of boloney about how great you were.

In the corporate world, we have it down to three levels: exceeds expectations, meets expectations, and does not meet expectations. This is simpler and more realistic, but it still leaves a lot of room for subjective interpretation and error. This person has these values, that person has those. This person judges by how things "look" from a distance, that person judges by how things actually are. The end product is what actually makes the money, but unfortunately a lot of people get caught up in the show of getting there.

A supervisor who gives a subjective evaluation is also given an evaluation from their superior, and so on. This of course creates a situation where people tend to look at the outside covering of people and the way they act, not how they are, and what really motivates them. And by the same token the outer manifestations of a problem are looked at, not the core causes of it. The symptoms, not the causes. And the more levels between the bottom of the organization and the top, the more muddled the communication becomes. Everyone is worried about how things look, not how they are, and problems follow. In a pyramid organization, a person cares about what their supervisor *thinks*, or their supervisor's perception of things, not actually about the work they are doing, or how efficient it is. This is why large companies tend towards inefficiency. The larger and older they become, the worse it gets. Only a free market in a democratic body can stay young and vibrant. None of this is actually news to a lot of people of course. For those of us who work in large corporations, we live it everyday. But the answer to the problem seems to elude most people for some strange reason. Again, people will always talk about how the system *should* work, if people would only follow it correctly.

The union company, however, tends to resemble the socialist state. It wants to treat everyone equal. The industrious and the lazy are the same; both enjoy the same pay and benefits. In the final analysis, the lazy worker gets even more. That is all well and good (if all companies were American union companies or European socialist companies), but in the long run the whole organization suffers if another company out there is competing in the same market and not using the union/socialist model. Then the union company loses and all the work goes to China.

The union companies are some of the oldest, most established, and higher paying jobs. But their environments often seem as a departure from reality. The union protects the workers from feeling that they have to work, and the management cannot easily address this, so a pathetic and juvenile tit-for-tat standoff ensues. The people look busy if necessary, while accomplishing nothing. In addition, they whine about this and that. The management feels their authority threatened, and instead of directly addressing the issues, they play junior-high hall-monitor games like requiring people adhere strictly to break times, the manager stands by the time clock at the end of the day, or at the beginning, to make sure no one is five minutes late, or leaves five minutes early, and various other little games. In addition, when work cannot proceed for one reason or another, often the workers at a union company will have to stay and try to look busy. A sense of value is completely absent. The lower level management might be fully a partner in this problem. A practical business atmosphere does not exist.

If people are not getting anything accomplished, it really doesn't matter if they take longer breaks or are ten minutes late. And if people *are* getting things done, it doesn't matter either. The only reason someone gets something done in a day is if they are motivated to do so, whether that motivation is positive or negative, internal or external. Ten minutes here and there might have economic value to a statistician, but in reality it means nothing and is just game playing, causing a general departure from reality in the workplace on both sides. Further compounding the problem is being at work when there is really nothing to do. If group "A" is held up because of a lack of progress by group "B," it is harder to see where the problems are when both groups are still at work looking busy. If group "A" went home, that would put a lot of pressure on group "B" to free up the work. Again, this culture promotes tremendous inefficiency, and a departure from reality. In addition, a two-way animosity exists, with the result that the moment the company doesn't need a group of people they lay them off, often times seeming out-of-the-blue and without rational to the workforce.

As Americans, we are used to living in a democracy and freely question authority. We are used to living by the *law,* not dictatorial authority. Somewhat anarchist or anti-authority. These two spheres of influence conflict our sense of order. Many large corporations operate under a written code which emulates the law of the land, but the corporation being a dictatorial monarchy tends to violate its own laws when it sees fit to do so, just as the monarchs of old did; for "business" needs. This is often understandable, but still leaves the individual in a quandary over which way to jump sometimes—to submit to authority or fight for one's perceived rights.

A company will firmly stress protecting its proprietary information to maintain its competitive advantage in the marketplace. The individual might then consider that they are protecting their own "proprietary" information, their expertise, to remain competitive in their own labor "marketplace." This does not help the company's goal of producing a product however, and is an impediment. Ideally, within the company information would be freely distributed in a spirit of teamwork towards the end of producing the product. Human nature would seem to tend contrary to this spirit of teamwork, however, and the selfish nature prevails—self interest.

There seems to be a double standard at play, which tends to hinder overall progress. The company wants to protect its proprietary information at all costs, while at the same time expecting its employees to be dutiful, selfless, and all giving. Of course many employees are not selfless and all giving so this creates inefficiency which slows the entire organization.

The types of people that are most skilled with their hands very often do not see far down the road, or the "big picture." They tend to work for the immediate reward They are practical about their surroundings, and take their world for its literal value; not how their superiors tell them they should look at their role. For example, if one is paid by the hour, one realizes the more hours the better. Overtime hours are worth more than regular hours; so saving work to be done on overtime hours is advantageous to the making of money for an hourly wage worker. This is fairly simple. This makes economic sense in the world of the hourly-wage worker. This is of course counterproductive to the production of a product. So some type of pressure must be applied to gain efficiency.

In the current corporate system, the "manager" is supposed to be the businessman. They are supposed to be working on a budget and be running a "business." But often it doesn't seem to work too well because it is somewhat "pretend" and the manager is often not very familiar with the actual work, and is therefore detached from the realities of the work environment. The manager is primarily focused on what their superiors think, the people who judge them and pay them, and less concerned with their actual product and customers. This relationship tends to be counterproductive in addressing the problems of the bottom, where the profits originate.

An example comes to mind. At Goodrich I had a friend who had a bad transmission in his old Toyota truck he needed to have repaired. He had first gone to a corporate chain transmission repair shop, and talked to the "manager." A manager is someone who is someone else's servant. A manager is a cog in the wheel. My friend asked this person for an estimate. After some deliberation, an estimate could not be given without first bringing his truck in.

Then my friend went to an independent transmission shop and talked to a business owner. While describing the transmission failure symptoms, the shop owner finished my friend's sentence as to what the transmission was doing, told him exactly what was wrong with the transmission, an gave him a precise estimate of the repair bill, all in one sentence.

Experience and observation has led to the following premise: A complex labor service can only be efficiently performed with quality and expeditiously completed by very small organizations working for a profit motive. The only possible exceptions to this are people working with an eye toward a "higher calling" such as a military or religious imperative.

I submit to you that management, i.e. the concept of *management*—to *manage* a business, at least in the context of a larger corporation, is a sometimes-necessary *evil.* It is a method of utilizing people and resources that should be minimized as much as possible. A more useful and practical method is the profit motive applied in the context of a business. This keeps things more "clean." The corporation should have a customer/provider relationship with its labor resources, not a pretend one, but a real one.

A world of management and managers results in a world far removed from a physical reality. We are all familiar with negative connotations to management—brown-noser, spineless jellyfish, two-faced, Napoleon, Hitler, etc. But, I would submit to you that these are not necessarily bad people, but *weak* people, shaped by the institutional framework they are immersed in. It takes an immense amount of good character not to succumb to the usual manager bad habits. And not all that many people these days have good character to begin with.

The businessman is *expected* to work hard for profit, it is accepted, even respected. The manager on the other hand, is a power hungry social climber, one that climbs on the backs of others, if they are aggressive. Maybe that's not fair, but is it true nonetheless? Motive and perception are the differences.

Lean attempts to introduce some marketplace concepts into the major industrial corporation, namely the dynamic relationship between provider and customer. But since the corporate authority and pay structure are the same, this means that to buy into the Lean ideals is to be somewhat gullible, or compliant. Of course what pushes forward the capitalist marketplace is the exact opposite characteristics of being gullible and compliant. So, Lean in an environment of well-educated and savvy people doesn't work as well. If people know their bread is buttered in such and such a fashion(x), they always have that(x) in at least the back of their mind, even if told to think otherwise.

Psychologists tell us that to hypnotize a person, you tell them to do one thing, or focus on one thing, while forcing them to focus on another thing. It seems that our world often tries to do just that.

What would work the best in America is a corporate structure that resembles our nation, a Democratic Republic, which operates via a capitalist economy. A work place that functions like an economy instead of an army; the Corporate Republic, operating via an *internal* capitalist economy. Just as the greater corporation functions in a capitalist economy, within the corporation a bunch of smaller capitalists are at play, each working for a "profit." A work place that is like a town or village; one that has some form of democratic organization, and with checks and balances. Instead of an army-like factory, a vibrant marketplace. A work place that has a lot of different entrepreneurs operating within it. That is the fire in the engine of progress.

Think of *Microsoft Windows*, utilizing a window inside a window, a solar system within a galaxy, a galaxy within a universe. Corporations compete on the world stage, one with the other. In the same way, in this system you would have an entire economy *within* the corporation, with all the resulting energy.

In the case of a firm that does large and complex projects, the efficiency of delivering the larger product is brought about by the individual small organizations seeking their individual profits. And, the individual contractor's profits have to be directly connected to the delivery of the larger product, thereby forcing them into cooperation, i.e., “we’re all in this together.” Like on a fishing boat.

It is not really all that far fetched of an idea, actually. Construction works this way. One construction company, the general or prime contractor, just handles paper and deals with the customers. They then subcontract out all the actual building functions, the iron, concrete, plumbing, electrical, etc. Many operations could be run in this fashion, but few are. We also have the example of the franchise, such as a fast food restaurant that is owned by a private individual, but is tied to the resources, and controls, of a larger corporation.

The basic building block of this system in the airline world would work, in my mind, like the tradesman system of old with the master, the journeyman, and the apprentice. The master would own the business in most cases, and in turn employ and train the journeymen and apprentices.

The Middle Age was an agrarian world and industry was in infancy. The guilds were said to have been extremely protectionist. The guilds regulated and controlled tradesman activities to a choking point. They were democratic associations much like what is being proposed here. But they were known to quash competition and the free market spirit of things. The guild system fell just before the industrial revolution, and was seen as counter productive to industry, creating stagnation, and a lack of innovation. So, you would not want to copy that aspect of them. It is possible to learn from the past. But, the small business utilizing the master assured excellent quality. Efficiency would come from competition. The guilds failed not because of the basic building block, the small business, but because of the distorted political and social problems that grew out of the political association that was the guild.

The army of Great Britain used to be supplied with muskets by hundreds of small gunsmith shops across England. They did not have parts standardization, but they did have recognized excellent quality of the product. Father-and-son shops, putting their personal best in every musket. This is the way American industry should go in this modern age. Technology makes it possible. Computers handle all the accounting.

Americans are independent minded. I am heavily imbued with Scandinavian blood, and we *really* don't like being told what to do. This kind of corporate structure would better fit the American character. Let the Asians have Lean.

Who have represented the majority of people who have invented and created things in our human history? The independent and free thinkers. The ancient near east was known for large empires in which everyone was a slave to the sovereign, rich and poor. Then we start to have some city-states that were more or less democratic here and there. In Athens they took knowledge from around the world and built on it, and created much. What distinguished them? They had a large middle class of basically free people. Those people impacted the whole world forever. In Europe, who did most of the creating? Noblemen, men of means, people with resources and time on their hands. They were as free as could be allowed at the time. Serfs do not create much, they do not "think out of the box." Who has done much of the creating in America? Just about anybody with a free spirit. Bicycle mechanics invented the aeroplane.

I used to know a guy who climbed to the tops of trees, and cut them down bit by bit. This was his own business, and he actually made a lot of money doing it. Like most loggers he was fiercely independent. In his spare time he designed and built offshore racing type tunnel hull boats. He had a greater knowledge of composite structures than anyone I have ever talked too, and yet he had no more than a high school education. But he was free, and it was his passion. However, you can't have freedom without some accountability.

The Corporate Republic would, in my mind, have the investors at the top as now, obviously. The investors would appoint a CEO as now. The CEO would appoint vice presidents, who would appoint directors, managers, and etc. The difference would be that at the level where direct labor management comes into play, the manager in direct charge of a group of labor would be subject to a "vote of confidence of the workers," or the group of subcontractor business owners. The manger would have his helpers, supervisors and work coordinators. A parallel would be like the town mayor with his staff & police. This manager would be accountable to both the upper management and to the group of subcontractors. He is appointed by upper management and paid by them, but he can be fired by a vote of no confidence from the subcontractors. The subcontractors are kept in line by being paid by the job like in any business. No results, no money. Real world.

The closest thing that comes to mind for an example, would be the chamber of commerce of a town with a common and central purpose, like a tourist town with a central theme, or a community that serves primarily one trade, such as ship building or the maritime repair industry, or a town exclusively involved in farming. Similar to the guild village of old that mostly made shoes or the like.

These subcontractors would be small companies, sometimes a one-man show, and probably not more than twenty-five people in a company. In general, not more than the owner can personally know. This seems to be key in a small business.

There could be democratic institutions (hey, what an idea?) like courts or boards of inquiry to handle disputes. For violations people would pay fines like traffic tickets.

When mistakes are made to the product and etc., insurance would come into play. The corporation pays the whole bill for a mistake now, but in this system it would be shared. Also, the subcontractors would bring some capital investments to the table like tools.

Yes, now you are working for a tyrant in some cases without union protection, but if you don't like him you can simply go work for this guy over here, right in the same building. Voila, competition for labor, what a concept. And, you can simply become a subcontractor yourself and be the owner, it would be attainable.

Using the airliner repair station as a model[6], let us take a look. The repair station consists of a hangar, a small engineering department with technical publications writers, a training department, a marketing department, groups of skilled labor with their supervision, a warehouse with a procurement department, a tool room, ground support equipment with personnel to maintain such, facilities maintainers, and an administration with the usual HR departments and so forth to oversee things. There is a substantial investment in real estate and equipment, however, the business is essentially a labor service. Without the skilled labor, there is for all practical purposes, nothing. The skill of the labor force has a great impact on the value of the product. You cannot simply

---

[6] See Appendix E, p. 200, A Case Study.

plug in people of a much lower skill level and get the same level of product.

The repair station has a book of operating procedures. The individual aircraft have their various manuals. These are all FAA approved and are akin to written down laws. In the real world, if you break the law and get caught, you pay a fine or spend time in jail. Enough said.

An airliner is one large project. It involves the work of a hundred people or so, which are divided into, say, five groups based on skill area. If the work is done by the proposed subcontractor units, let's say the work is done by fifteen or twenty subcontractors. On the corporation side, you have work coordinators that the subcontractors interface with. There would have to be some system for determining who gets to do what. Also on the corporation's side are inspectors to "buy" the work on behalf of the company. The inspectors are accountable to the company and FAA, pretty much the same as now. The company would provide insurance for mistakes. The subcontractors would be financially accountable here, and if they make a lot of mistakes, their insurance would of course go up. If they were really bad, they would perhaps be completely driven out of business.

There is a tool room involved, a warehouse, and heavy ground equipment. These operations would best be run as businesses in their own right, with very real customers. We have effectively a tool store, and tool rental store, a heavy equipment rental business, a hard parts store, and a hardware store. An engineering services business. A restaurant instead of a cafeteria, and etc.

The mechanics already own their own tools, their "capital" investment. Now they would own a lot more tooling, as the company would provide almost nothing. What it does supply, it would rent to the subcontractors.

The parts and supplies are already paid for by the airline customer, so this part of the equation is fairly easy. Just a matter of keeping everyone honest, and that the parts and materials are billed correctly.

The work package of the airplane is divided up into hundreds of individual jobs. These jobs range from hundreds of man-hours to just a few. This is the way it is now, so this part of the equation is an easy jump to working for a profit "by the job." Pretty simple really, in basic concept. Not rocket science. In reality it would take a much more involved and complicated form, like society does. But, this is very doable, and as we see in the construction industry example, already in effect.

Possibly a likely form would be a turnkey franchise business, developed by the larger corporation, one designed with success in mind for all. Onc that takes into account the democratic and independent spirit of Americans, and puts these ideals into form.

If you think that all this is ridiculous, you might want to ask yourself what the alternatives are. Instead of Big Brother watching us, he seems to be asleep at the wheel, some might even say he's lost his head up his backside.

In the non-union company, you work for unappreciative tyrants, and the workers are unhappy. Traditional union company, the work goes to China, and the workers are unhappy. Maybe, just maybe, it's time to try the American way.

Let's think of a corporation as a wild cannibalistic beast, a predator, and as part of a species of like beasts, such as an industry of similar corporations. The species (industry) could be manufacturers of aircraft, say. Individual beasts would be x, y, and z corporations. In our era of environmental awareness, we know that a wild beast must be free in the wild to thrive. The beast lives in the "market" if you will. It is a predator competing in a market for resources (market share). This keeps it strong and healthy—the species thrives in this environment. If a beast becomes weak or dies, its flesh is eaten by the other beasts and provides sustenance for the rest of the species, and the species lives on.

If there is only a few beasts left, say, because they have eaten each other down to a few, and they are taken into captivity—they become fat, unhealthy and lethargic. They fail to thrive and die, causing an extinction. The carcass is left to rot, not providing sustenance for the others of the species as before—for there is no more. All is wasted. This would be like the bulk of an industry being taken over by government "for its own good" and "managed." The result is likely extinction.

Corporations need labor, and therefore provide jobs. If a corporation within an industry fails, the workers can go to another corporation with that industry rather easily. The workers really work for the industry, more so than an individual company. If the entire industry fails, everybody is out of luck.

So what would the religious reason be for this being a "mission from God" anyway? Let's look at some Bible facts and then do a recap of subjects talked about while stepping way back and look from the macro perspective. Lets look at some history as a function machine where a given input x produces an output y.

We learn in the Bible that Satan has the lion's share portion of control over the world's resources and politics. But while God is infinite, Satan is quite finite. Satan has a very limited capacity in celestial terms. Satan can only work his best through a state with an unquestioned central authority like Napoleon or Hitler. Such a state has to be built just at the right time, just under the right circumstances. In the example of Nazi Germany, anarchy, crisis and poverty were the order of the day; that needed to be "fixed" by just the right man with the plan. Just give all control to him and he will fix you up.

We learn we will have Satan's antichrist in the end times, and this person will have control of the earth. How could one person gain control of the earth in practical terms? The earth would most likely have to already have the means to be ruled from a central point. This was impossible in ancient times due to the distances involved and natural barriers such as mountain ranges and oceans. Such physical barriers are no longer a problem with the Internet, mass communications, and modern air, sea, and land travel, not to mention the space arena.

What of the political barriers to one world government and control? Free market democracy seems to be a barrier to such a scheme. So how could Satan manipulate world systems to get rid of such uncontrolled democracy? —By giving it a disease known as socialism to weaken it until it collapses, where dictators and emperors take over.

Europe was a Christian dominated land, at least on the surface, for millennia. European religion, culture and political structure changed little from the fall of Rome, the Dark Ages, until the 1400-1500s, when it started to progress some and explore the world. During all this time it is based on monarchy and a plethora of feudal kingdoms. The Catholic Church was a strong central political force that presided over all of Europe until the Reformation in the 1600s, after which Christendom was left with multiple churches.

Trailing the Reformation in Europe, the intellectual movement known as the Enlightenment spans the 1700s.

The Enlightenment would be heralded as reason and science over the traditional sway of the church and state.

Democracy is born into the modern world in the form of the United States, 1776. The United States is born as a Christian nation, based on that ethic, and taking its governmental forms from ancient Greece and republican Rome. The only official "king" the United States has is God Himself, and this is not by accident. In the same year, 1776, Adam Smith publishes his book *An Inquiry into the Nature and Causes of the Wealth of Nations,* the seminal work of capitalism. These new dynamic forces, democracy and capitalism, set in motion changes in the world. The western world grew even more prosperous.

Following the American Revolution in Christian Europe, the ideals of the Enlightenment, hubris and large-scale war would be the impetus for change in the form of the French Revolution, the Terror, and Napoleon.

In the large business concerns created in the wake of capitalism, rich business magnates grow to represent unfeeling monarchs of the feudal era. Workers increasingly feel slighted. In this period atheist intellectuals founded the concepts of the various brands of socialism, with Karl Marx as their icon with his *Communist Manifesto*, coauthored by Friedrich Engels. In the Manifesto Engels states that communism "...is destined to do for history what Darwin's theory has done for biology...." Atheist minds think alike.

In the *Manifesto*, Communism is oriented at the wage earning working class, socialism more at the middle class. The communists would work to take over the state through revolution, and they would work to take over industry through labor unions sparked by their labor organization, the IWA, wielding the strike as their revolutionary power. Both means brought violence and destruction. Communism seeks to eliminate the proprietor, the capitalist businessman (bourgeoisie) who employs wage labor (proletariat), and replace them with state control. Marx even defends feudalism. The stated aim of the *Manifesto* is a new world order.

Special but uneducated everyday workers take up the cause with fervor, though many of them are Christians still. They fight their cause with passion. They feel the pain of mistreatment. Eventually, they begin to score victories with business and government.

In 1914 the advent of World War I was the great turn for the worse in Europe, the great futile war of pride, followed by its re-flare known as WWII, after which the European dominance over the world was quite at an end. At this point Europe was now predominantly socialist and atheist, and of waning economic and political importance. The enemy's work was done there.

That is the state level result of European nationalistic pride teamed with socialism and an absence of God—defeat. But what of the work of the communist IWA in the unions at the industrial level, and of the economic prosperity brought by industry in a capitalist framework? In the late twentieth century this ongoing fight starts to take its toll. Corporations and whole industries fall victim to the continuous struggle and die. In America, the steamship industry dies. Manufacturing industries die. The end of the labor movement struggle seems to be "everybody loses." Is this an unintended consequence? Some industries deemed necessary to the government are kept alive by the government—such as shipyards for the Navy. The trend is that the government must take over and run things. The central power. The pattern for one world government…?

The ironic thing is that socialism is proven to be ineffective, yet the trend is to embrace it still by atheistic pro-government control types. Monarchy also has proven ineffective. Democracy and the market have proven the best performers, though not all the time thus giving critics a reason to complain. There are examples going back through history where a loosely controlled body of traders and/or manufacturers with a degree of freedom to trade as they will has made a prosperous city such as ancient Tyre or medieval Venice. The market is God's answer for His purposes. Namely, He controls it and no one human person does—the "invisible hand" in Adam Smith's work. The market is too large and complex for Satan to thwart effectively. The market goes up and down. That is the nature of things. You save in the prosperous times for use in the lean times. It has purpose.

We are still engaged in that fight, we are still learning the lessons. The industry will shape the politics, not the other way around. Where one works is the "real" world, the day-to-day experience. If we embrace the capitalist free market in a democracy we can hold off the enemy a little longer.

Appendix A

This is a letter to the president of BfGoodrich Aerospace, aircraft maintenance and overhaul division. (1997)

TO: John Martin

FR: Chris Parker, mechanic / MLSP

RE: MRO Profitability

I have for some time now had some ideas floating around in my head and I feel compelled to share them with you. I believe that there is a way to make this division one of the most efficient of its kind in the world. I am a college student with major aviation maintenance management. I also believe in God, and I like to study history. This combination has led me to the following points of view.

In the past, when people have been trying to make improvements at Tramco, they seem to not take psychology into account. They fail to understand the nature of the problem. You're dealing with people here, not machines or products. Statistical thinking alone will not work here. The company ever since I have been here has always had a leaderless feel to it, and the people have felt they were being used to make someone else money. I am not judging here, merely stating fact. Things are much better now, but we have run up against a wall, efficiency wise.

Appendix A

By my life's experience and my learning, I have seen that there are only two leadership styles that work. Harsh and oppressive using fear, and dutiful magnanimous giving leaders. I happen to think the later works much better. The "middle of the road" path is very ineffective. In psychology it is taught that to achieve a result, both positive and negative stimuli work, but that positive actually works much better.

I have been maintaining aircraft most of my adult life now and I know people at airlines, Boeing, etc. I think that our whole aviation maintenance culture is flawed. For all practical purposes, the jet age is only one generation old, it is still very new from a historical point of view. The maintenance aspect of this jet business has fairly humble origins. All the emphasis up to this point has been on the engineering, conquering the great scientific milestones. Also, the pilots who fly these aircraft have received a great amount of attention. On the other hand, maintenance is viewed as a necessary evil to keep the profit machine rolling. It is no wonder that we have evolved the way we have.

I read a lot of history, I view myself as kind of an amateur anthropologist. It is my belief that you can compare an American corporation to a form of government, or a city-state. I think that what works in the organization and government of people in general, would work at the corporate level as well. So, since most of us will agree that democracy is what works best in terms of form of government; I think that is what works best in terms of the management of a corporation. Actually, the democratic republic idea which is our America is what most suits the corporate idea. I think this thinking follows along the lines of Total Quality Management as well. The hard part from the management perspective is giving up control. We all like control, to rule, but we see from history that control doesn't work well.

From my study of history, I have learned that the more

democratic a society is, the more rich it tends to be. The more a society's citizens are free thinkers and doers, the more creative and productive they are. When I read about the various ancient civilizations, I find that most in the ancient world are very similar. By the ancient world I mean most civilizations before the time of Christ. They are always characteristic of a strong central government and at the head of the government is of course the strong or god-like king. Everyone in this society is basically a slave. Most willingly stay in their slave positions within their country because there is security in their land. But, in this context, creativity is not what it could be and progress is fairly slow or nonexistent. I could go on and on but you get the picture.

Then comes ancient Greece. Ancient Greece is painted by historians as being in a category all by itself. Somehow, they develop what is known to us today as the democratic society. Now they did still have slaves, but we will just have to give them a break here. Their citizens were not slaves. They were free to think and do as they choose. This, of course, was a very prosperous land indeed as history has shown. It has been shown that the ancient Greeks very, very nearly invented the steam engine. If they had, it is assumed that they would have had our present day technology thousands of years ago, as a result of the ensuing chain of events. What a score for the democratic society! Of course, the Greeks were more interested in a good time and the pure pursuit of art and philosophy than the practical applications of science.

Then came Rome. Rome conquered Greece and in turn took in a lot of Greek philosophy. This helped Rome stay on top for many centuries, but they were still control freaks to some extent, and they slowly became corrupted, and the barbarians came and got them. But before Rome fell, it became Christian, and there were no more slaves, at least technically. The barbarians of northern Europe got God as well and they chilled out to a great extent, but then they developed the feudal system. Now the people in this system were not technically slaves, but many of them might as well have been. The Baron owned all the land, and he let people use it for a nominal fee. Right back to our highly centralized little governments like in the ancient oriental slave world. To me, the example of the feudal manor is much like the modern corporation. It has been shown, by the example of America vs. Europe, that this is not a productive system.

So, out of the misery of this feudal system, some guys decide to come up with communism, the opposite extreme. The IAM is a communist organization in my opinion. Well, we all know where communism leads-- poverty. Ruin the economic system so bad that it falls in on itself. No one has incentive to be productive in the communist state or organization. All people are not equal as workers, some are more motivated than others, some need more leadership than others. But if the unmotivated and the motivated get paid the same, eventually all are unmotivated. That is what I think the IAM is doing to the aircraft manufacture and maintenance industries. It is creating a system so inefficient that we will not be able to compete globally, maintenance wise.

The way I think a company such as ours should be treated is like a village. It should be kind of like home. People should work on the planes as if they owned them their selves. We spend more of our waking lives at work than elsewhere, so there is no good reason that it should be a bad experience. After all, we are all in this together, and the people are what is making the company its money.

There are examples of villages or cities which became very rich as a result of collective effort, but individual interests. On the Mediterranean you had ancient Tyre of Phoenicia and then much later Venice of Italy which were trading and manufacturing cities. As cities they were very rich and influential, but they were made up of individual interests. We need to always take what works from history and continually take it a step further.

I believe in God; a lot. So did the people who started the United States. My direct ancestors in Massachusetts were among these. I can tell you they really had faith. They believed in business, each man owning his own property, and fairness. When the Mayflower pilgrims came over, their first settlement was like a hippie commune, where everybody shared everything. Well, nobody liked that, it did not work well at all. So, they divided up all the land and gave each family a plot to farm, but it was still understood that theirs was *one* community. The men served in *one* militia. They all went to *one* church. They clearly derived benefit from being part of that one community. Amish people live very similar to this day. Each has his own farm, and yet they are still one community, they help each other out for the mutual gain. One man's barn burns, everyone in the community builds it back. No problem. This is much better than the *modern* American ideal of every man for himself. Every man for himself and we all eventually lose. No, we can't have the philosophy of every man for himself. We must admit there is prosperity in numbers--just not when one man takes power and enslaves all the rest. There must be democracy. Free enterprise.

When I was unemployed a few years ago, I started a business repairing boats and cars. It does a funny thing to you, starting a business. It makes you see things you did not see before. It really makes you see that the limiting factors that an entrepreneurial person has are self-imposed ones, you're free. But it also shows you how difficult it is to do everything by yourself. You kind of need partners if you are really going to make a go of it. Again; safety and prosperity in numbers. So, I have come to the conclusion that the group situation is best, and that is what led me back to aviation.

We have to some how get the spirit of free enterprise and democracy on the hangar floor, for real. The only one who can really initiate that is you. I have a few ideas on what can be done toward this goal.

Appendix A

I.

Change how we are paid. The concept of the wage provides no incentive to do anything but be at work a certain amount of hours. Like it or not that is the way it is in the real world. I'm here, I know. My idea is to do this: If a man makes $15/hr, then pay him $10/hr to be here, and give him the other $5 in the form of a percentage of the company's earnings for that month/quarter. Two thirds of his pay for his individual effort, and *one third* of his pay is derived from the *group effort* of the company, I think two thirds/one third are good proportions. Eventually, it will sink in that when the company makes more money the people make more. Once this gets going, it will be like setting a fire. Additionally, if a mechanic needs tools to do this work, its fair that he gets a yearly tool allowance. Like a company expense account. This will make the average guy feel like a real big shot. He will feel just a little more like a professional. It would be money well spent.

I look at a IAM airline like United, and I see a place where the average guy who has been there a while makes about $25/hr. One size fits all based on seniority. Again, communist. Through people I have known, I know that in the 1980's, the United work force was very fat and a large percentage of the people just sat around and did nothing all day, every day. Well, this can't be too good in the long run. Its acceptable in the fat years when no one wants to make waves. But then, lean years come and layoffs, and everyone's in a fit. The people are laid off based on seniority. What if the work force was based on $35-$40/hr. aircraft gurus, and $10-$15/hr. helpers, with education and training respective of both. Would this make layoff time easier, while maintaining a quality work force? To me, this would be more "real world", people have differing motivations, and seniority number doesn't always have a lot to do with it. This type of system also gives a junior guy more to shoot for. A better reason to be industrious.

II.

We can improve the quality of our product. Since our product is the work of people, it makes sense that we need to improve the quality of the worker, the overall quality of our people, both moral and technical. This needs to be done across the whole American aviation maintenance world. It *could* happen here because there is no IAM to stand in the way. We could develop to the point where we *define* how aviation maintenance is performed. If we did that, do you think companies would pay good money to get work done here? I think so. It has been shown that quality wins in the long run.

For people to be productive, they must be motivated and forward moving. Stagnation brings about decay, there is no treading water. Our current system is fairly stagnant. We have engineers upstairs in their world and mechanics on the floor in their world. They should be closer together. Some of the lower engineering functions I'm sure could be better handled from the hangar floor where the perspective is better. At the same time it would be very beneficial for at least a segment of the mechanics to be more schooled in the engineering point of view. I personally have seen my own efficiency improve as I have taken math courses for example, plus I try to learn engineering terms and concepts. This stuff does make a difference.

I like to use the medical model. You have doctors, who have a lot of school, hands on training, and make a lot of money. Then you have the RN's, the LPN's, and so on down the line. We should be a little more like that so there can be a progression, as well as a high top standard. Professionals and helpers. That is how many career fields operate.

We should foster a career ideal, put people on a career path of constant progression. I think aviation maintenance should be divided into two categories based on the psychology involved in doing the jobs: systems and structures. These should be the two main aviation maintenance career paths. Avionics, systems, and flight controls all involve schematic thinking and moving parts, and in the future, computers. Structures is kind of a craft that involves a feel for the subject as well as an idea for the strength of materials and types of repairs.

Right now it seems that in our company, at our pace, it takes about two years to become reasonably fluent in one area, whether it be systems, flight controls, or structures. I don't know if that applies to avionics, but I do know that much of the avionics jobs are not all that hard for an intelligent person to pick up and that you don't really need all the math and electronics training that some people think.

The current A & P school system ranges from a joke in one extreme, to not all that applicable to the technician of transport category aircraft in the best extreme. We should do something to augment the traditional A & P school's in the form of an apprenticeship.
Waiting for the government to do something is not a wise idea, we will be waiting forever. I realize we are doing an A & P program here, but what I'm talking about is a greatly modified curriculum. One that goes in depth into the SRM, OHM, MM, WDM, and corporate standards, etc., of transport aircraft.

So, this is how I think an aircraft technician should start his/her career:

FAA AMT(T): Focus mainly on systems. Two years flight control, two years systems, two years avionics.

FAA ARS: Focus on structures. Structures, structural bond, NDI, repair engineering.

If you would like to hear more, please let me know.

Sincerely,

Christian Parker

Appendix B

To: Claudia (Director of United's Oakland, Ca.
maintenance base) (June-July 1997)

FR: Chris Parker, new hire, bay 3 sheetmetal

RE: The maintenance world culture

I am a new hire to Bay 3 sheet metal shop. I have worked on aircraft most of my adult life and I have about eleven years of heavy turbine aircraft experience. I spent eight years in the Air Force right near here at Travis AFB, where I got to work on the NASA airborne observatory, and I spent a few years at BfGoodrich Aerospace in Everett, Wa. At BfG I worked on aircraft from Southwest Airlines, Alaska, Western Pacific, Sun Country, UPS, FeDEX, Ansett Australia, Casino Express, America West, etc., as well as several corporate jets and the President of Mexico's personal 757. Additionally, I have known many people who worked for United in the past when I was in the Air Force, and I know mechanics at Alaska and Northwest Airlines. I think I have seen a variety of the aircraft maintenance world.

At BfG, the major customers especially, and even the minor customers to some extent, have their aircraft maintained just the way they want, that is to say, they get to more or less use their own maintenance system and paper work. For example, everyone working on United aircraft had to go to a special United paperwork class, and the United parts system is used. Southwest has their aircraft maintained at BfG just as if they were doing it themselves, as well as UPS, FeDEX, Western

Pacific. America West gets their aircraft maintained much better than they themselves do it.

I had some great expectations of United coming in the door. I expected to see an operation run superior to the repair station that I came from. I wanted to feel like I had made a good choice for my future, I thought the employee owned thing was the ticket for the future. I did not think I was going to see anyone working at a feverish pace, knowing about the IAM world for years, but I did expect to find people who new what they were doing and had broad based knowledge. I expected to find a good training system and education assistance like they have at Boeing.

Well, so far I am surprised to say that I am not very impressed. I realize that I have only been here a few weeks and I am at the tail end of a heavy check, and there is not much work to be done. But the general attitudes I run into here I find very childish and unprofessional. I am pleased to hear that the management wants to change things, but I see a primitive state on the floor. Year of experience for year of experience, a person from BfG is vastly superior to a typical United worker on first impression. The clear reason is the system. This atmosphere smells of welfare. Few places do people show ambition. What good does it do for a person to show up on time for work if they just spend their day "looking busy". They tell me I can be fired for being late from break, or calling in sick, or taking a pen home, but I ask you, what are they going to say when this whole operation gets farmed out due to inefficiency? I can go back to work at BfG for the same money I'm making today--in a matter of days. I'm not so sure the people I'm around would find the same transition easy.

To me, its not a matter of working hard or easy. It's a matter of attitude, trying to get things accomplished or just trying to look busy. In this kind of work, if you know what you are doing, you can look like you're doing very little, yet get much accomplished in reality. Or, you can not care about what you are doing, not care if you actually get anything accomplished, but look busy or stay out of sight and make people think you are doing good.

Clearly what is need here is a moral and philosophical shift, not just mere policing and tattling. One of the problems I see here though, is many of the workers born and raised here do not see any problem. They think they are good to go, and that everywhere else has a problem. But as for me, I feel like I am back working for the government, and that is not saying much.

I am not trying to be purely negative here, I would very much like to see this place prosper if I am going to be here, but I feel something should be done soon. I have attached a letter I wrote to the manager of BfGoodrich MRO division and his subsequent reply. Interestingly I wrote this letter just a couple of weeks before being called by United for an interview. This letter more fully expresses my attitudes towards my profession.

Of course I realize that I could be fired for even stating these ideas around here as a probationary employee, but I just feel like I can't wait six months or longer. I have a strong faith in Almighty God and somehow this stuff just seems to come out of me rather uncontrollably.

If I can be of any further assistance please feel free to contact me regarding this matter.

Appendix C

To: John Martin, President, BFG Airframe Services
Division (1997-8?)

Fr: Christian Parker, Master Mechanic ML11

Re: suggestions for business

Hello, it has been a long time since I last wrote you a letter. Do you remember me? I have been wanting to write you again concerning how I think this business should be conducted based on my experience here on the floor. I am a religious person with points of view that are usually very eccentric to my fellow workers. What I write in the next few lines will probably make you think I'm nuts, but I am going to tell it like I see it anyway.

There are times when I actually think that I am on a mission from God to get rid of the IAM union. I think that to achieve this I need to present a method for doing business at this company which both prospers this company, and all the labor workers that are employed here. The workers at the IAM companies will then see that there is a better way than their union and they will want to adopt our system, and all America will be the better for it. Yes, it's true, I hold these delusions of grandeur. Or at least from time to time.

Appendix C

People who work at IAM and big union companies tend to believe that the union is a necessary evil. Yes, many of them think the union is an evil, but better than the alternative--a company like us. They look at a non-union company and they see less pay and benefits (typically) and they say "see, I told you so." And in most cases they are right, but in our case good old competition has closed the gap considerably here at BFG and other repair stations.

But still, the big union companies are the ones to compare wages too, they are still the leader in most people's eyes. If the unions are ever to be truly defeated, something better has to be offered. Something that ordinary people can readily see the fruits of.

And from your point of view, the corporate BFG point of view, you want to see this company make more money, and that's the bottom line--that's business. You represent the interests that control the capital, and they want results. But the interests are also people like you and I, we are investors in BFG and would like to see the company do well too. It's a public company and no one person really owns it or controls it. The people who run it are supposed to govern it toward the greater good, or they will be replaced. And round we go. The problem is that at some level the company ceases to become a with a board of directors and becomes little monarchies and fiefdoms.
What I think is needed in the workplace here and elsewhere is a strange and obscure concept: Capitalism, and Democracy. Now most people would say that it should be obvious that we are in a capitalist system here, free country and all, a capitalist country. But when I look around me, and analyze my place at a big company, I don't feel free or capitalist. It feels more like being a soldier in an army or a peasant in a feudal society, where the lord owns and controls everything. These are the problems the Europeans dealt with centuries ago and that the United States supposedly solved in practice.

Aviation historically is a very, very managed business. It has to be because aviation safety is at stake people say; you can't just let any old fool do whatever he wants to do to these airplanes, he might do the wrong thing and kill everybody. There is also the fact that a big aerospace company like this requires substantial capital, and if I put out a group of capital I want some control over how it is used, right? Or do I just want to make money? I say it has to be one way or the other. I say less *management* equals more money. But, you can't just give people more pay and be nice to them and expect the greatest results. You have to be nice to them in a different way; you have to give them control over their own destinies.
democratic body

I like to study history, it fascinates me. I do not have nearly the knowledge of it yet that I would like to have, but what I do have leads me to a few conclusions which I will talk about shortly. First let me say one thing; I think that what is needed to get what BFG wants out of this company is already here. I think that management here has tried almost everything, but nothing seems to quite do the trick. We have gotten better, but we are still quite a ways from being a well oiled machine. I think we are at the place of change where instead of making the ship go faster, it's time to invent the airplane. It's time for a quantum leap where labor relations are concerned. The problem with such leaps is that they are often leaps of faith, with many doubters.

I have come to see a corporation such as ours as like a state, or a feudal manor. I do not believe I am the originator of this concept, but I have come to see the implications of it on my own. We are like a monarchy, we have a king--you. I think you are a good king, but you are a king nonetheless in this organization; maybe a better analogy would be a general. No one can make any meaningful change here unless you approve it. And it extends down like that to me. Just like in the military. Is the military an economically viable business? Not according to most people. But we always want to model our corporations after the military. And I believe that is why so often corporations do not get the productivity out of labor that they would like.

If a corporation is like a state, then what kinds of states historically perform the best? Monarchies? Communist states? Democratic Republics? Monarchies have always been at least somewhat limited in wealth and power, or short lived. Communism was a backlash to a monarchy. The Russian Tzar was overthrown by the Bolsheviks because the Russian people were so unhappy with monarchy. I believe this is how unions formed. The union was a backlash to abuses of power in feudal like companies. The people revolted economically. Most of the union organizers of the 1930's were Marxist. These were philosophical beliefs they had. Philosophical beliefs can apparently be very powerful. Does communism work? I think the answer is obvious. Does IAM labor work? I think the answer is obvious. What about Democratic Republics? The Greeks of Athens, as an almost pure democracy, are said to have been right on the verge of discovering modern technology 2500 years ago. Rome, the greatest country the world had ever known until now, grew to be the Rome at the top of the world as a republic for it's first 500 years. Then it switched to being an empire, or monarchy, and after a brief peaking out dwindled down to nothing and fell apart. You could say their stock took a long tumble. The republic of Rome was the model for our government of the United States that has been talked so much about. We all seem to feel that it has done pretty well. It would appear that democracy works best. So why do we model our corporations after monarchies or military units? When it comes to economics, democracy and capitalism are clearly best.

One reason that capitalism works better than a run-from-the-top, pyramid structure that I have come to see is the idea of macro/micro. You are one man, you can only handle so much stuff in one day, you delegate other stuff to your subordinates, and they in turn, etc. That organization functions like a flow chart, a series of blocks, all nice and neat and simple. To me that is the macro level of the organization. Then comes my version of the micro. The micro level of the organization is the bizillion little transactions that happen down here on the hangar floor. That, I am convinced, is where the money here is made or lost. Leads, supervisors, etc. can not keep track of all these things without really slowing things down. If the individual worker does not take care to keep the ball rolling, no amount of managing will help. But what incentive does the individual worker have to keep the ball rolling? Personal integrity? These days? Fear of reprisal in one form or another is what it is in our monarchy right now, mostly. A negative incentive. As we established in a former letter, positive incentives work better. A positive incentive would be a worker getting paid on the actual jobs they do, and only getting paid when the whole job is done, i.e., airplane delivery. Getting paid for doing something, as opposed for getting paid to not be disappointing. That would provide the incentive for the worker to grease the machine and keep the ball rolling. To put their heart into it, like, for real. As opposed to getting paid by the hour for just being here, which for all too many provides little incentive to produce.

The analogy of macro/micro that has come to my mind is that of the human body. I can understand the workings of the body at the anatomical level. As a mechanic I see it acting much like a machine, the muscles, tendons, bones, circulatory system, nervous system. A lot like the parts in a jet these days. That's like the macro level. But then there is the physiological/microbiological level that totally loses me. How could anyone possibly keep track of all that stuff if it weren't automatic? I do not think any amount of management would help, it's just to much. When they try to interfere with it they usually just mess it up. See my comparison?

Simply put, *management* is inefficient. That is not to say that leadership is not needed, it very much is needed, but *management* is less efficient than people working for their own interests, for *profit*. The profit motive is a greater incentive than the urge to keep-out-of-trouble or not-be-seen-in-disfavor motive.

So how would we make our company function as a capitalistic democracy? To me, it is as simple as each person being paid for what they actually do, and the workers having a vote in the events that effect them and the people that govern them. Each maintenance line might be considered a “village” of workers with elected leaders and representatives, etc., even while the company still has it’s supervisors or “agents” making sure the company’s needs are being met. Maybe even a constitution and the rule of law for the labor body. Take out a lot of subjectivism. A jury of peers for a strict disciplinary action.. This stuff would not work well in an environment where people are paid by the hour to be here, but in an environment where people are only paid on what they produce, this would work well. It's not so easy to "play the system," when if you do not produce, you do not make much money. We all win or we all lose. This is the general idea. Do this and we will all be rich. Of course it is easy to see that the actual implementation will be a little more tricky, and I have a starting plan outlined below.

Here's my big plan in a nutshell. I will lay it out and then justify each of my statements further on.

- I think that the mechanics on the hangar floor should be in business for themselves. That every licensed mechanic should be his or her own subcontractor outfit, with mechanics teeming up where necessary in small groups. This is already provided for in the Repair Station Manual(RSM), see class V subcontractor. An example would be Newell or Tank Tigers.

- Unlicensed mechanics would work directly for licensed ones, but everyone would be paid based on their actual customer billed hours, say $25 to $35+ dollars per customer billed hour, depending on whether they are licensed or not and whatever other criteria you would like. Pay would come at the close of a job number, so as soon as the plane leaves all people on that plane would be paid.

- I think that most transactions on the hangar floor should be real buy and sell transactions. Not -esque, not pseudo, not reminiscent.

- Violations of quality, or safety, or work progress would be dealt with utilizing small fines, like traffic tickets. It's all about money here, it's a business, so let penalties be monetary and fit the crime. Keep things in their proper perspective. Financially reward the desired behaviors and financially discourage the undesired behaviors.

- Mechanics would buy standards from the tool room, and then bill them to the customer. If they have some left over, they would keep them in their tool box properly marked, or sell them back to the tool room.

- Mechanics would be responsible for buying most of their tooling, this would be the mechanic's "capital investment," no more trying to find ways to keep the tool room properly stocked with serviceable tools. Tooling and standards are in the realm of the mechanic, the micro stuff, we are the ones who work closely with them, and if we are paying for them ourselves we would be the best judge of their proper disposition. The company would only be responsible for the really big, special, and expensive stuff, the macro stuff.

- Run many of the business functions here as businesses in their own right. The tool room would be open to outside customers, the training department as well, even the cafeteria.

- The company would provide or arrange insurance for mistakes and accidents (R-22 type situations).

- Many things would be virtually the same as now. You would still have the employment office, but now it would be used to screen people as to their suitability to work on site. You might want to sign contracts with mechanics for so many hours of maintenance per year, etc. You would still need QC, and even more so. QC would be accepting the maintenance from the mechanics on behalf of BFG. Same as it is now, only more so.

Now before you dismiss this all as foolishness or impossibility, listen to my argument.

I recently looked up capitalism in my encyclopedia and was intrigued at the similarity of what I think is needed at this company and the article on capitalism. This guy Adam Smith is credited as being the "father" of capitalism. He wrote a book titled "An Inquiry into the Nature and Causes of the Wealth of Nations(1776)." The following is an excerpt from the Microsoft Encarta 97 encyclopedia.

> Smith sought to show how it was possible to pursue private gain in ways that would further not just the interests of the individual but those of society[BFG & employees] as a whole.
>
> Society's interests are met by maximum production of the things that people want. In a now famous phrase, Smith said that the combination of self interest[pay by cust. bill hour], private property [personal tools & equip. i.e. physical means to do the job], and competition among sellers[workers?] in markets will lead producers "as by an invisible hand" to an end they did

not intend, namely, the well being of society.

Capitalism has certain key characteristics. First basic production facilities--land and
capital-are privately owned. Capital in this sense means the buildings, machines, and other
equipment used to produce goods and services that are ultimately consumed.

Second, economic activity is organized and coordinated through the interaction of buyers and
sellers (or producers) in markets.

Third owners of land and capital as well as the workers they employ are free to pursue their
own self-interests in seeking maximum gain from the use of their resources and labor in
production. Consumers are free to spend their incomes in ways that they believe will yield
the greatest satisfaction. This principle, called consumer sovereignty, reflects the idea that
under capitalism producers will be forced by competition to use their resources in ways that
will best satisfy the wants of consumers. *Self interest* and the pursuit of gain lead them to do
this[my italics].

Fourth, under this system a minimum of government[BFG] supervision is required; if competition is present, economic activity will be self-regulating. Government will be
necessary only to protect society from foreign attack, uphold the rights of private property,
and guarantee contracts.

Human beings, Adam Smith said, have always had a propensity to "truck, barter, and
exchange one thing for another."

If we apply these concepts to the inner workings of a corporation, they should yield the same results as America's prosperity. Companies compete with each other on the market, we need to bring the situation down one level so we have the same thing going on within the company as without. If there was ever a company that could actually do this, BFG Airframe Services is it.

Appendix D

To: Bill Ashworth (2007)

Fr: Chris Parker, mechanic, SWA

Re: State of the business

Hello Mr. Ashworth, I am a master mechanic on the Southwest unit. I have been known to write a letter to the Company President from time to time. I don't know if you have ever read one of them, but it wouldn't surprise me if you have. I have written to the last three, now four, CEO's. I am writing now for three reasons: I have a start date at Boeing, this business seems to be failing, and I didn't want to leave without trying one last time. I would like to see an entirely new aviation maintenance industry.

A corporation functions just like an army. A pyramid organization with a general at the top, and then down through the ranks to the private. The corporation is democratic at the very top, the shareholders, but then it becomes an army from the CEO down to the lowest labor level. In an army, the lower people answer to the upper people. In the end, this creates a lot of misinformation and inefficiency. Armies make good consumers, but are not known for their money making abilities. In the army, we pay people first, then we set standards and expectations. If we feel that the standards and expectations are not met, then we complain and discipline. Mostly negative.

Appendix D

I would like to see a workplace that operates as an economy--as an economy in a free republic. In the democratic republic, the leaders answer to the people. We have a parallel to this concept in the Japanese corporation. If the Japanese corporation does bad, the boss apologizes to the people.

Fundamental to the republic is business; lots of businesses competing with each other. Businesses try to please customers with their product or service. The owner of a business is not a servant, as the soldier in the army, and therefore is more self-assured. He's out there "trying to make a buck" by any means possible. This is psychologically more positive. When approaching a business for a service, you usually pay for the entire service when it is done. You do not pay someone by the hour to chip away.

Just a week ago, I was talking with one of my co-workers about the transmission he needed to have repaired in his old Toyota truck. He had first gone to a corporate chain transmission repair shop, and talked to the "manager." A manager is someone who is someone else's servant. A manager is a cog in the wheel. My friend asked this person for an estimate. After some deliberation, an estimate could not be given without first bringing his truck in.

Then my friend went to an independent transmission shop and talked to a business owner. While describing the transmission failure symptoms, the shop owner finished my friends sentence as to what the transmission was doing, told him exactly what was wrong with the transmission, an gave him a precise estimate of the repair bill in one sentence.
This example highlights the problems of a corporation that does service work. Is it possible to combine the best of the small business and the major corporation? I think it is possible, and have stated so in many letters to your previous bosses. But I think that, maybe, I did not come across to clearly before.

When I think of a model for this entity, what comes to mind, as an example, is a town that is involved in one activity. For example, Leavenworth, Washington, is a tourist town. It is comprised of many small businesses with a common theme—tourism, in Bavarian theme village setting. The local government is part of this scheme. But instead of the town being owned by a corporation that regulates everything to one master's wish, the town is a collection of small businesses and the city government is responsible to those citizens and businesses.
In a similar fashion, I would like to see an aviation maintenance organization here that comprises many small subcontractors, Goodrich as the prime subcontractor, and some kind of democratic organization, having a constitution, to provide checks and balances.
I am not talking about a traditional union company, which in the end functions like the socialist state.

I have thought of a lot of the details, but you might have read them already in a previous letter. So, I will close with this for now. I hope you will want to hear more. If not, I will go and play the Boeing and IAM game like everyone else.

I would expect that a man close to retirement would like to have a good legacy to look back on in his retirement years.

Sincerely,

Christian Parker

***

The author had a visit with the Goodrich president the very day the author left Goodrich to be a flightline AMT for the Boeing 787 program. The president did not see that there was any problem at all, even though the majority of his experienced workforce, the people that made Goodrich go, had gone over to Boeing and a number of other employment opportunities. He told the author that things at Goodrich were never better and that the company was making money. He was dead serious.

Two months later, the Air Force contract that Goodrich had hoped to land went to Boeing instead. A couple of weeks later the Goodrich ATS facility was formally announced to be sold to an Australian firm.

Appendix E

A Case Study

One fall afternoon when the author was hiking to the top of Mount Ellinor in the Olympic Mountains of Washington, I met a guy who was up hiking with his two boys. We got to talking, and it turns out that he was a harbor pilot, regularly piloting the ships from Port Angeles to Puget Sound.

I told him what I did and that my recent ancestors were marine engineers, and that that is how my family came to live in Washington. I told him that I lived up near Port Townsend. The conversation turned back to ships and boats. He made the observation that the shipwright people in Port Townsend were the best to be found anywhere, in his opinion. And if fact in the opinion of many, the quality of services received in the Port Townsend Shipyard are generally well received, and highly regarded.

The Port Townsend Shipyard is not run by Big Marine Inc., but rather a collection of smaller businesses belonging to a loose federation known as the Marine Trades Association. Many of these businesses only perform a specialized task and therefore are interdependent on each other. Some are larger and provide a broader spectrum of services, and are more oriented towards the very large boats and small ships the port services.

The Port Townsend boatyards are particularly known for their prowess with wooden boats, which require a high degree of skill to build and make professional repairs. The local shipwrights are equally skilled with metal construction and composites, as well as all the systems which comprise a small ship or boat.

What sets these marine professionals apart is that in most cases they are doing this work because that is what they most want to do, for the love of it, especially for the wooden boats. There is not a lot of money in marine repair these days, and efficiency is necessary to stay in business.

Every boat or ship is a unique project. No two are alike. There is not a standard template. The work of one boat to the next is often similar, but never exactly the same. Ingenuity on the fly is a must. You must rely on the skill and knowledge of the workers more than on a set process.

The Port of Port Townsend owns the marina and land adjacent to the marina, the entire shipyard complex. The port provides haul-out Travelifts for both the fishing boats and yachts, and a large one for small ships well over 100' long. The Travelift is what makes this kind of shipyard feasible. It is a giant wheeled lifting hoist that literally lifts the boats right out of the water, and then drives them wherever they need to go. Using a Travelift, there is no need for a boat to sit in a dry-dock, be constricted to a marine railway, or on a tidal grid at the beach. It is lifted out of the water on a special ways, driven to a parking spot on the port property, and set down on shoring so that it can be easily accessed for work.

The port leases space and buildings to the port's many varied businesses. Boat services, suppliers, equipment rental outfits, cafes, even other industrial businesses such as a small brewery and a coffee roaster.

The port supervises the safety and environmental compliance of the work being performed. Each boat project pays a rental fee to the port for the square-foot space the project takes, for the duration of the stay.

The federal government is represented on site by the Coast Guard, which is the equivalent to the FAA for the maritime industry.

There are a couple of large service providers, having engineering services close at hand in addition to labor, which are capable of building very large boats, or providing complete overhaul services. The largest of these generally employs less than a hundred people. These facilities make use of large "hangars" just like aviation hangars. In fact, the boats on their stands and shoring are very much like large aircraft on jacks, and shored, in hangars, surrounded by stands and man-lifts in similar fashion. One might compare the outside shipyard to the apron in front of an aircraft hangar. Variations of the same thing.

There are several smaller service providers, which take on projects of varying size. Again, they utilize "hangars," or more frequently do the work outside over tarps--to catch the environmentally hazardous debris.

The general service providers are in a position to co-ordinate with the specialized service providers and material & equipment providers to satisfy the needs of their customers.

One general mid-sized service provider that caters to the older wooden boats is the Port Townsend Shipwrights Co-Op. This business enjoys a stellar reputation. A Cooperative is of course a form of corporation in which the members own the business jointly and equally, and run the business in a democratic fashion. This Co-Op still hires other workers for wage labor on individual projects.

In the shipyard there are fabricators, which can make assemblies, or entire aluminum boats. There are specialized providers of every kind, including electrical services, marine canvas, marine engines, welding, machining, rigging, hydraulic installation, propellers, refrigeration, engineering services, marine surveyors, and etc.

There are service provider businesses that are truck based and do not have a facility on site, the same as in building construction.

Rather than one large corporation which attempts to do everything, the labor and engineering services and the warehousing of parts and supplies are separate businesses. One business is a marine service provider, another business sells hardware and parts, and maybe rents tools and equipment. Another business sells more specialized items such as engines, batteries, or propellers. Most businesses have a narrow focus; do one thing and do it well.

There is no company cafeteria here, but a few cafes who specifically target the boat workers, and alternately the people visiting the marina. And of course the café is accessible to the general public, not being housed within a private facility, and therefore can be more profitable throughout the day.

Bibliography

Adams, John. *Ocean Steamers: A History of Ocean-going Passenger Steamships 1820-1970* London: New Cavendish books, 1993.

Angas, W. Mack. *Rivalry on the Atlantic* New York, New York: Lee Furman, Inc., 1939

Annin, Robert Edwards. *Ocean Shipping: Elements of Practical Steamship Operation* New York, New York: The Century Company, 1920.

Armbuster, Kurt E. *Orphan Road* Pullman, Washington: Washington State University Press, 1999.

Barnes, Charles Brinton. *The Longshoremen* New York, New York: Survey Associates, Inc., 1915.

Benson, Richard M. *Steamships and Motorships of the West Coast* Seattle, Washington: Superior Publishing Company, 1968.

Best, Gerald M. *Ships and Narrow Gauge Rails* Berkeley, California: Howell-North Books, 1964.

Bixby, William. *Track of the Bear* New York: David McKay Company, Inc., 1965.

Buchanan, Joseph R. *The Story of a Labor Agitator* New York, New York: The Outlook Company, 1903.

Camfield, Thomas. *Port Townsend: An Illustrated History of Shanghaiing, Shipwrecks, Soiled Doves and Sundry Souls* Port Townsend, Washington: Ah Tom Publishing, Inc., 2000.

Camfield, Thomas. *Port Townsend: The City That Whiskey Built* Port Townsend, Washington: Ah Tom Publishing, Inc., 2002.

Canfield, George L, and Dalzell, George W. *The Law Of The Sea* New York, New York: D. Appleton & Company, 1921.

Canney, Donald L. *U.S. Coast Guard and Revenue Cutters, 1790-1935* Annapolis, Maryland: Naval Institute Press, 1995.

Carmichael, A.W. *Practical Ship Production* New York, New York: McGraw-Hill Book Company, Inc., 1919.

Bibliography

Chadwick, F.E., et. al. *Ocean Steamships: A Popular Account Of Their Construction Development, Management And Appliances* New York, New York: Charles Scribner's Sons, 1891.

Chase, Don M. and Ardencraig, Marjorie N.H. *Pack Saddles and Rolling Wheels: The Story of Travel and Transportation in Southern Oregon and Northern California from 1852* Grants Pass, Oregon, 1959.

Clark, Norman H. *Mill Town* Seattle, Washington: University of Washington Press, 1970.

Churchill, Winston S. *A History of the English-speaking peoples* New York, New York: Dorset Press, 1956.

Dear, Ian. *Great Ocean Liners: Heyday of Luxury Travel* London: B.T. Batsford, Ltd., 1991.

Dewing, Arthur S., PhD. *Corporate Promotions and Reorganizations* London: Harvard University Press, 1914.

Dolin, Eric Jay. *Fur, Fortune, and Empire* New York, New York: W.W. Norton & Company, 2010.

Dollar, Robert. *Memoirs of Robert Dollar* San Francisco, California: Robert Dollar Company, 1925.

Eells, Rev. Myron. *The Twana, Chemakum, and Klallam Indians of Washington Territory* Washington, District of Columbia: Smithsonian Institution, 1889.

Epstein, Steven A. *Wage Labor and Guilds in Medieval Europe* Chapel Hill, North Carolina: The University of North Carolina press, 1991.

Evans, Holdan A. *Cost Keeping And Scientific Management* New York, New York: McGraw-Hill Book Company, 1911.

Floherty, John J. *High, Wide and Deep* Philadelphia, Pennsylvania: J.B. Lippincott Company, 1952.

Forbes, B.C. *Men Who Are Making The West* New York, New York: B.C. Forbes Publishing Co., 1923.

Foster, George H., and Weiglin, Peter C. *Splendor Sailed the Sound* San Mateo,

California: Potentials Group, Inc., 1989.

Fowler, Chuck, and Withers, Dan. *Patrol And Rescue Boats On Puget Sound* Charleston, South Carolina: Arcadia Publishing, 2011.

Gibbon, Edward. *The Decline and Fall of the Roman Empire* Garden City, New York: Nelson Doubleday, Inc., 1963.

Gibbs, James A. *Pacific Graveyard* Portland, Oregon: Binfords & Mort, Publishers, 1950.

Gibbs, James A. *Shipwrecks off Juan de Fuca* Portland, Oregon: Binfords & Mort, Publishers, 1968.

Gibbs, Jim. *West Coast Windjammers in Story and Pictures* Seattle, Washington: Superior Publishing Company, 1968.

Hansen, Clas Broder. *Passenger Liners from Germany: 1816-1990* West Chester, Pennsylvania: Schiffler Publishing, Ltd., 1991.

Haskell, Burnette G. *What The I.W.A. Is* San Francisco, California, 1881.

Hawkins, N. *Maxims and Instructions for The Boiler Room* New York, New York: Theo, Audel & Co., Publishers, 1902.

Hawkins, N. *New Catechism of The Steam Engine* New York, New York: Theo, Audel & Co., Publishers, 1903.

Haugland, Marylou McMahon. *A History of Alaska Steamship Company* Seattle, Washington: University of Washington Press, 1968.

Herm, Gerhard. *The Phoenicians: The Purple Empire of the Ancient World* New York, New York: William Morrow and Company, Inc., 1975.

Hines, J.S. *Pacific Marine Review* San Francisco, California: 1918.

Howden, J. R. *The Boys' Book of Steamships* New York, New York: Frederick A. Stokes Company, Publishers, 1911.

Howe, R.S. *The Great Northern Country: Being the chronicles of the Happy Travelers Club in their pilgrimage across the American continent as traversed by the Great Northern Railway Line and Northern Steamship Co. from Buffalo to the Pacific Coast* St. Paul, Minnesota: Great Northern

Railway and Northern Steamship Co., c. 1895.

Huebner, Grover G. *Ocean Steamship Traffic Management* New York, New York: D. Appleton and Company, 1920.

Huebner, Stephen Solomon. *Marine Insurance* New York, New York: D. Appleton and Company, 1920.

Huenecke, Klause. *Jet engines* Osceola, Wisconsin: Motorbooks International Publishers & Wholesalers, 1997

Hunn, Peter. *Beautiful Outboards* Marblehead, Massachusetts: Devereux Books, 2002.

Hunn, Peter. *The Old Outboard Book* Camden, Maine: International Marine, 2002.

Irving, Washington. *Three western narratives: a tour on the prairies; Astoria; The adventures of Captain Bonneville* New York, New York: Literary Classics of the United States: Distributed by Penguin Books, 2004.

Japanese Mail Steamship Co. (N.Y.K.). *To Nippon, The Land Of The Rising Sun* Sydney, Australia: John Andrew & Co., Printers, 1899.

Jones, Gwyn. *A History of the Vikings* Oxford, England: Oxford University Press, 1984.

Jordan, Roger W. *The World's Merchant Fleets 1939* Annapolis, Maryland: Naval Institute Press, 1999.

Kinnison, Harry A. *Aviation Maintenance Management* New York: McGraw-Hill Inc., 2004

Lane, Frederic C. *Ships for Victory: A History of Shipbuilding under the U.S. Maritime Commission in World War II* Baltimore, Maryland: The Johns Hopkins University Press, 1951.

Lawson, Will. *Pacific Steamers* Glasgow: Brown, Son & Ferguson, Ltd., 1927.

Lien, Carsten. *Exploring The Olympic Mountains: Accounts of the Earliest Expeditions 1878-1890* Seattle, Washington: The Mountaineers Books, 2001.

Liker, Jeffrey K./Hoseus, Michael. *Toyota Culture* New York: McGraw-Hill

Inc., 2008

Loong, Michael. *Essentials of Airplane Maintenance* BookSurge, LLC, 2005

MacElwee, Roy S. *Ports and Terminal Facilities* New York, New York: McGraw-Hill Book Company, Inc., 1918.

MacElwee, Roy S., and Taylor, Thomas R. *Wharf Management, Stevedoring and Storage* New York, New York: D. Appleton and Company, 1921.

Macfarlane, Robert. *History of Propellers and Steam Navigation* New York, New York: George P. Putnam, 1851.

Malone, Vincent J. *The Story of the Marine Firemen's Union* San Francisco, California: Pacific Coast Marine Firemen, Oilers, Watertenders and Wipers Association, 1945.

Marx, Karl. *Das kapital : a critque of political economy* Seattle, Washington: Pacific Publishing Studio, 2010.

Marx, Karl and Engels, Freidrich. *The Communist Manifesto* New York, New York: SoHo Books, 2013.

Matthews, Frederick C. *American Merchant Ships 1850-1900* New York, New York: Dover Publications, Inc., 1987.

McCurdy, James G. *By Juan de Fuca's Strait* Portland, Oregon: Binfords & Mort, Publishers, 1937.

McDonald, Lucile Saunders. *Alaska Steam: a pictorial history of the Alaska Steamship Company* Anchorage, Alaska: Alaska Geographic Society, 1984.

Miller, Byron S. *Sail, Steam and Splendor: A Picture History of Life Aboard the Transatlantic Liners* New York, New York: Times Books, 1977.

Miller, William H., jr. *The Great Luxury Liners 1927-1954* New York, New York: Dover Publications, Inc., 1981.

Newell, Gordon R. *Ocean Liners of the 20th Century* New York, New York: Bonanza Books, 1963.

Newell, Gordon R. *Pacific Coastal Liners* Seattle, Washington: The Superior

Publishing Company, 1959.

Newell, Gordon R. *Pacific Lumber Ships* Seattle, Washington: The Superior Publishing Company, 1960.

Newell, Gordon R., and Williamson, Joe. *Pacific Tugboats* Seattle, Washington: Superior Publishing Company, 1957.

Newell, Gordon R. *SOS North Pacific* Portland, Oregon: Binfords & Mort, Publishers, 1955.

Newell, Gordon R. *Ships of the Inland Sea* Portland, Oregon: Binfords & Mort, Publishers, 1960.

Newell, Gordon R. *The Green Years* Seattle, Washington: Superior Publishing Company, 1969.

Newell, Gordon R. *The H.W. McCurdy Marine History of the Pacific Northwest* Seattle, Washington: The Superior Publishing Company, 1966.

Niven, John. *The American President Lines and its Forebears 1848-1984* Newark, New Jersey: University of Delaware Press, 1987.

Nystrom, J.W. *A Treatise on Screw Propellers and their Steam Engines* Philadelphia, Pennsylvania: Henry Carey Baird, 1852.

Osbourne, Alan. *Modern Marine Engineer's Manual* New York, New York: Cornell Maritime Press, 1941.

Pacific Coast Steamship Co. *All About Alaska* San Francisco, California: Pacific Coast Steamship Company, 1890.

Pacific Mail Steamship Co. *A Sketch Of The New Route To China And Japan* San Francisco, California: Turnbull & Smith, Book and Job Printers, 1867.

Parker, Rev. Theodore. *Genealogy and Biographical notes of John Parker of Lexington* Worcester, Massachusetts: Press of Charles Hamilton, 1893.

Perry, Fredi. *Seabeck: Tide's Out. Table's Set.* Thompson Falls, Montana: Perry Publishing, 2005.

Powers, Dennis M. *Tales of the Seven Seas: The Escapades of Captain Dynamite Johnny O'Brien* Lanham, Maryland: Taylor Trade Publishing,

2010.

Radnoti, George. *Profit Strategies for Air Transportation* New York: McGraw-Hill Inc., 2002

Riegel, Robert. *Merchant Vessels* New York, New York: D. Appleton and Company, 1921.

Rodengen, Jeffrey L. *Evinrude-Johnson, and The Legend of OMC* Ft. Lauderdale, Florida: Write Stuff Syndicate, Inc., 1993.

Sale, Roger. *Seattle, Past to Present* Seattle, Washington: University of Washington Press, 1976.

Sautner, Randolph J. *American Mail Line, Limited 1926-1974* Oakland, California: American President Lines, Claims Department, 1993.

Schwantes, Carlos A. *Railroad Signatures across the Pacific Northwest* Seattle, Washington: University of Washington Press, 1993.

Schwantes, Carlos A. *Long Day's Journey: The Steamboat & Stagecoach Era in the Northern West* Seattle, Washington: University of Washington Press, 1999.

Schwartz, Stephen. *Brotherhood of the Sea: A History of the Sailors' Union of the Pacific* San Francisco, California: Sailors' Union of the Pacific, 1986.

Smith, Adam. *An inquiry into the nature and causes of the wealth of nations* Dublin: Printed for Messrs. Whitestone, Chamberlaine, W. Watson, Potts, S. Watson, 1776.

Spears, John R. *The Story of the American Merchant Marine* New York, New York: The MacMillan Company, 1910.

Stark, Peter. *Astoria* New York, New York: Harper-Collins Publishers, 2015.

Stine, Thomas O. *Scandinavians on the Pacific, Puget Sound* Seattle, Washington, 1900.

Strobridge, Truman R. and Noble, Dennis L. *Alaska and the U.S. Revenue Cutter Service 1867-1915* Annapolis, Maryland: Naval Institute Press, 1999.

Talbot, Frederick A. *Steamship Conquest of the World* Philadelphia, Pennsylvania: J.B. Lippincott Company, 1913.

Tate, E. Mowbray. *Transpacific Steam* Cranbury, New Jersey: Rosemount Publishing and Printing Corporation, 1986.

Thiesen, William H. *Industrializing American Shipbuilding: The Transformation of Ship Design and Construction, 1820-1920* Gainesville, Florida: University Press of Florida, 2006.

Thompson, Wilbur, and Beach, Allen. *Steamer to Tacoma* Bainbridge Island, Washington: Driftwood Press, 1963.

Turner, Robert D. *The Pacific Empresses: An illustrated history of Canadian Pacific Railway's Empress liners on the Pacific Ocean* Victoria, British Columbia: Sono Nis Press, 1981.

U.S. Department of Commerce, Coast and Geodetic Survey. *Alaska supplement to Coast Pilot Notes on Bering Sea and Artic Ocean.* Washington, D.C. Government Printing Office, 1911.

U.S. Department of Commerce. Coast and Geodetic Survey. *Annual Report of the Superintendent of the Coast and Geodetic Survey.* Washington, DC. Government Printing Office, 1908-1932.

U.S. Department of Commerce, Coast and Geodetic Survey. *General Instructions for the Field Work of the Coast and Geodetic Survey.* Washington, D.C. Government Printing Office, 1908.

U.S. Treasury Department, United States Coast Guard, *Record of Movements: Vessels of the United States Coast Guard 1790 – December 31, 1933.* Washington, DC. Coast Guard Historian's Office, 1989.

U.S. Treasury Department, United States Revenue Cutter Service, *Report of the Cruise of the U.S. Revenue Cutter Bear and the Overland Expedition for the Relief of the Whalers in the Artic Ocean from November 27, 1897, to September 13, 1898.* Washington, D.C. Government Printing Office, 1899.

Villiers, Capt. Alan. *Men, Ships, and the Sea* Washington, D.C: The National Geographic Society, 1973.

Waterhouse, Frank. *Pacific Ports Annual* Seattle, Washington: Pacific Ports,

Inc., 1919.

Waterman, T.T. *The Whaling Equipment of the Makah Indians* Seattle, Washington: University of Washington Press, 1920.

Waterman, T.T., and Coffin, Geraldine. *Types of Canoes on Puget Sound* New York, New York: Museum Of The American Indian Heye Foundation, 1920.

Waterman, T.T. and Greiner, Ruth. *Indian Houses of Puget Sound* New York, New York: Museum Of The American Indian Heye Foundation, 1921.

Wead, Frank. *Gales, Ice and Men* New York, New York: Dodd, Mead & Company, 1937.

Wellington, John L. *The Gold Fields of Alaska* Cripple Creek, Colorado: The Buckner Printing Company, 1896.

Weintraub, Hyman. *Andrew Furuseth: Emancipator of the Seamen* Berkeley, California: University of California Press, 1959.

Wood, Robert L. *The Land That Slept Late. The Olympic Mountains in Legend and History* Seattle, Washington: The Mountaineers, 1995.

Index